Calanoid Copepods of the Genus *Scolecithricella* From Antarctic and Subantarctic Waters

Taisoo Park

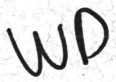

Paper 2 in

Biology of the Antarctic Seas IX
Antarctic Research Series Volume 31

Louis S. Kornicker, Editor

American Geophysical Union

CALANOID COPEPODS OF THE GENUS SCOLECITHRICELLA FROM ANTARCTIC AND SUBANTARCTIC WATERS

TAISOO PARK

BIOLOGY OF THE ANTARCTIC SEAS IX
Antarctic Research Series Volume 31

Edited by LOUIS S. KORNICKER

Copyright © 1980 by the American Geophysical Union
2000 Florida Avenue, N. W.
Washington, D. C. 20009

Library of Congress Cataloging in Publication Data

Park, Taisoo
  Calanoid copepods of the genus Scolecithricella
from Antarctic and subantarctic waters.

  (Biology of the Antarctic seas; 9, paper 2)
(Antarctic research series; v. 31)
  Bibliography: p.
  1. Scolecithricella—Classification.
2. Crustacea—Classification. 3. Crustacea—
Antarctic regions—Classification. 4. Eltanin
(Ship). I. Title. II. Series. III. Series:
American Geophysical Union. Antarctic research
series; v. 31.
QH95.58.B56 vol. 9, paper 2 [QL444.C72]
ISBN 0-87590-151-4   574.92'4s [595'.34] 78-31901

Published by the
AMERICAN GEOPHYSICAL UNION
With the aid of grant DPP77-21859 from the
National Science Foundation
April 11, 1980

Printed by
THE WILLIAM BYRD PRESS, INC.
Richmond, Virginia

CALANOID COPEPODS OF THE GENUS *SCOLECITHRICELLA* FROM ANTARCTIC AND SUBANTARCTIC WATERS

TAISOO PARK

Department of Marine Sciences, Texas A&M University, Galveston, Texas 77550

Seventy-four Isaacs-Kidd midwater trawl and 66 Bongo plankton net samples from antarctic and subantarctic waters have been examined for the calanoid copepod genus *Scolecithricella*. A total of 5284 adult copepods representing 19 species of *Scolecithricella* were found. Six species are new to science, 2 species (*S. profunda* and *S. obtusifrons*) are new records to the area, and 3 species represented solely by males are not identified. Antarctic copepods previously referred to *S. glacialis* and *S. polaris* are recognized as belonging to *S. minor* and *S. emarginata*, respectively. All species are characterized with pertinent descriptions and illustrations. Keys are presented for identification of the species.

INTRODUCTION

A large number of midwater trawl samples have been collected in antarctic and subantarctic waters on the USNS *Eltanin* under the U.S. Antarctic Research Program and processed at the Smithsonian Oceanographic Sorting Center. Some of these samples, most of which were previously studied for the calanoid copepod families Aetideidae and Euchaetidae [Park, 1978], and Bongo plankton net samples taken in antarctic and subantarctic waters on USNS *Eltanin* cruise 46 have been examined for the families Phaennidae and Scolecithricidae. During the examination a number of species referable to the genus *Scolecithricella* of Scolecithricidae were found. They have turned out to be extremely difficult to identify, mainly because a large number of species have been described for the genus and many of them have not been adequately characterized. As was noted by Vervoort [1951], Bradford [1973], and Roe [1975], the family Scolecithricidae has not been properly classified, nor have its genera been clearly defined. It is commonly found in the literature that closely related species are placed in different genera and certain unrelated species are brought together in one genus. It is therefore impossible at present to determine exactly how many species have been

described that are properly referable to the genus *Scolecithricella*.

The present study was intended to characterize all species of *Scolecithricella* found in antarctic and subantarctic waters in terms of complete anatomical and distributional analysis and to find reliable characters useful in classifying the species into groups. Since most of the *Scolecithricella* species are deep living and rare, they are seldom caught in sufficient numbers for adequate studies in conventional plankton net samples. The samples for the present study were therefore selected from the large midwater trawl collections taken obliquely between the surface and deep depths and Bongo plankton net samples obtained obliquely or horizontally from various depth ranges.

During the present study, 74 midwater trawl and 66 Bongo plankton net samples taken widely in antarctic and subantarctic waters were examined, and a total of 5284 copepods (4711 females and 573 males) referable to *Scolecithricella* were found. The females were identified with 16 species, and the males with 10 species. Three of the 10 male species were not equated with any of the female species. In view of the number of samples examined and the extent of the area covered, it is believed that the species found in the present study represent nearly

25

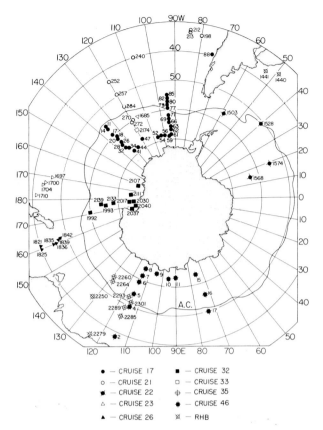

● — CRUISE 17   ■ — CRUISE 32
○ — CRUISE 21   □ — CRUISE 33
✳ — CRUISE 22   ⊕ — CRUISE 35
△ — CRUISE 23   ✱ — CRUISE 46
▲ — CRUISE 26   ⊠ — RHB

Fig. 1. Locations of collection stations. Station RHB 1434 is not shown. A.C. is Antarctic Convergence.

the entire fauna of the genus *Scolecithricella* in the antarctic and subantarctic seas.

MATERIALS AND METHODS

The 74 Isaacs-Kidd midwater trawl and 66 Bongo plankton net samples on which the present study was based were all selected from USNS *Eltanin* collections of the U.S. Antarctic Research Program except for three midwater trawl samples, which were taken from the South Atlantic by Richard H. Backus on R/V *Atlantis II* cruise 31. The 71 *Eltanin* midwater trawl samples were from cruises 17 (28 samples), 21 (9 samples), 22 (4 samples), 23 (5 samples), 26 (6 samples), 32 (10 samples), 33 (1 sample), and 35 (8 samples). The 4 samples from cruise 22 were taken in the Atlantic sector, the 8 samples from cruise 35 in the Australian sector, and the remaining 59 samples in the Pacific sector, including areas off New Zealand (Figure 1). The 66 Bongo plankton net samples were collected in the eastern half of the Indian Ocean

sector on *Eltanin* cruise 46 and represent 33 tows because the paired Bongo nets collected two samples in each tow.

All midwater trawl samples were collected by means of a double-oblique tow, in which the trawl collected all the way from the surface to a desired depth and again to the surface. The Bongo net tows were made either double obliquely between the surface and a desired depth or horizontally at a specific depth by means of an opening-closing device operated by messengers. The depth of sampling in both the midwater trawl and the Bongo net tow was determined from the wire angle and the length of wire out. The cod end of the midwater trawl was either 366- or 569-μm mesh net. The net mesh sizes used for the Bongo nets were 366 and 505 μm. The collection data, including the date, local time, position, and depth, are shown in Table 1.

Only a portion of each trawl sample was examined to pick up the adult copepods, and the volume of sample examined varied from sample to sample as well as from species to species. However, all Bongo net samples were examined completely except for certain shallow-water samples that were excessively large.

All measurements of specimens were made under a dissecting microscope with an ocular micrometer. The prosome length is the distance between the anterior end of the forehead and the posterior end of the metasome as measured from the side. The urosome length is the distance from the anterior margin of the genital segment to the distal end of the caudal ramus as measured in dorsal view. The body length is the distance from the anterior end of the forehead to the distal end of the caudal ramus when the specimen lies straight on its side.

The measurements were usually made after staining and clearing the specimen in lactic acid, but specimens that did not require a close examination for identification were measured in a 5% formalin solution, the same preservative as was originally used for the sample. Anatomical studies were made with specimens stained with methyl blue and cleared in lactic acid. All figures of the body and appendages were drawn with the aid of a Wild drawing tube attached to a Wild M20 microscope. Anatomical terms used herein have been defined by Park [1978].

Genus *Scolecithricella* Sars, 1902

The genus *Scolecithricella* was erected by Sars [1902] for *Scolecithrix minor* Brady, 1883, which was

TABLE 1. Collection Data

Station	Date	Local Time	Position Start	Finish	Collecting Depth, m
			Eltanin Cruise 17		
11	March 24, 1965	2110–2215	55°54′S, 135°06′W	55°55′S, 135°13′W	73–0
14	March 26, 1965	2100–2205	56°53′S, 135°02′W	56°53′S, 135°02′W	265–0
17	March 28, 1965	1557–1628	59°01′S, 135°33′W	59°02′S, 135°34′W	219–0
18	March 29, 1965	1357–1445	60°54′S, 134°26′W	60°53′S, 134°26′W	768–0
20	March 30, 1965	1540–1645	61°56′S, 135°12′W	61°56′S, 135°12′W	768–0
26	April 1, 1965	1326–1700	63°48′S, 135°02′W	63°57′S, 135°06′W	2560–0
28	April 2, 1965	0432–0454	64°04′S, 134°57′W		183–0
32	April 3, 1965	1100–1155	65°57′S, 135°00′W	66°00′S, 134°56′W	411–0
34	April 4, 1965	1118–1215	66°56′S, 134°00′W	66°59′S, 135°00′W	183–0
41	April 7, 1965	1634–1756	68°01′S, 130°58′W	68°02′S, 130°46′W	625–0
44	April 8, 1965	1432–1540	68°03′S, 127°20′W	68°02′S, 127°00′W	625–0
47	April 10, 1965	1757–1914	66°57′S, 120°10′W	66°57′S, 120°10′W	457–0
52	April 13, 1965	1707–1835	67°57′S, 106°46′W	67°57′S, 106°48′W	1052–0
54	April 13, 1965	1335–1550	67°55′S, 103°13′W	67°55′S, 103°04′W	684–0
56	April 14, 1965	1503–1650	67°59′S, 99°43′W	67°55′S, 99°01′W	1251–0
59	April 16, 1965	1723–2000	68°40′S, 95°35′W	68°36′S, 95°40′W	1251–0
61	April 17, 1965	2055–2250	68°05′S, 95°00′W	68°01′S, 95°00′W	1251–0
62	April 18, 1965	1801–2010	67°01′S, 94°40′W	66°55′S, 94°35′W	1251–0
63	April 18, 1965	1515–1715	66°08′S, 94°21′W	66°06′S, 94°21′W	1251–0
66	April 20, 1965	0840–0955	64°13′S, 95°01′W	64°09′S, 94°59′W	311–0
69	April 21, 1965	1330–1625	62°56′S, 95°15′W	63°01′S, 95°22′W	3146–0
71	April 22, 1965	1440–1606	62°06′S, 94°44′W	62°04′S, 94°04′W	457–0
77	April 25, 1965	1632–1758	59°00′S, 95°17′W	59°01′S, 95°19′W	836–0
79	April 27, 1965	0008–0120	58°00′S, 95°00′W	57°57′S, 95°03′W	386–0
80	April 28, 1965	0000–0135	56°57′S, 95°16′W	56°57′S, 95°14′W	625–0
82	April 29, 1965	0105–0224	55°56′S, 95°02′W	55°57′S, 95°01′W	313–0
85	April 30, 1965	0028–0140	55°02′S, 94°54′W	54°59′S, 94°57′W	625–0
88	May 9, 1965	1922–2340	39°24′S, 74°58′W		2502–0
			Eltanin Cruise 21		
198	Nov. 26, 1965	0435	34°00′S, 80°36′W	34°11′S, 80°46′W	2972–0
212	Nov. 28, 1965	1612	33°01′S, 83°53′W	33°06′S, 83°57′W	500–0
213	Nov. 28, 1965	1835	33°06′S, 83°57′W	33°13′S, 84°01′W	1050–0
240	Dec. 11, 1965	0040	39°54′S, 107°36′W	39°56′S, 107°40′W	2470–0
252	Dec. 16, 1965	2258	44°03′S, 120°17′W	44°10′S, 120°22′W	1050–0
257	Dec. 19, 1965	0309	49°10′S, 120°15′W	49°19′S, 120°19′W	1000–0
264	Dec. 21, 1965	0715	54°17′S, 119°46′W	54°26′S, 119°44′W	1230–0
270	Dec. 26, 1965	1454	60°17′S, 119°54′W	60°18′S, 119°58′W	128–0
272	Dec. 26, 1965	1745	60°21′S, 120°03′W	60°24′S, 120°09′W	1000–0
			Eltanin Cruise 22		
1503	Jan. 23, 1966	0510–0830	57°04′S, 59°33′W	56°59′S, 59°13′W	2505–0
1528	Feb. 5, 1966	0346–0719	50°54′S, 40°15′W	50°43′S, 40°20′W	2452–0
1568	Feb. 23, 1966	0827–1240	63°05′S, 15°02′W	63°02′S, 14°36′W	2359–0
1574	March 2, 1966	1739–2039	55°20′S, 18°55′W	55°11′S, 19°01′W	2608–0
			Eltanin Cruise 23		
1685	May 5, 1966	0301–0637	59°08′S, 115°14′W	59°11′S, 115°44′W	2250–0
1697	May 19–20, 1966	2236–0522	46°40′S, 170°03′W	46°22′S, 170°14′W	2274–0
1700	May 21, 1966	1355–1816	44°37′S, 172°58′W	44°21′S, 172°48′W	1275–0
1704	May 22, 1966	1336–1546	43°50′S, 174°27′W	43°45′S, 174°19′W	800–0
1710	May 25, 1966	1658–1956	41°44′S, 178°18′W	41°45′S, 178°05′W	900–0
			Eltanin Cruise 26		
1821	Dec. 4, 1966	1534–2203	40°07′S, 161°08′E	40°03′S, 160°41′E	3150–0
1825	Dec. 5, 1966	0218–0613	39°58′S, 160°34′E	39°49′S, 160°29′E	1625–0
1835	Dec. 11, 1966	0438–0916	45°19′S, 160°11′E	45°27′S, 160°17′E	1375–0
1836	Dec. 11, 1966	0829–1222	45°29′S, 160°12′E	45°38′S, 160°12′E	2181–0
1839	Dec. 12, 1966	1536–2205	47°00′S, 162°00′E	47°16′S, 161°55′E	3750–0
1842	Dec. 13, 1966	0230–0613	47°28′S, 161°52′E	47°35′S, 161°48′E	1350–0
			Eltanin Cruise 32		
1992	Jan. 3, 1968	1327–1816	60°04′S, 170°53′E	60°12′S, 170°33′E	3660–0
1993	Jan. 5, 1968	1247–1519	65°58′S, 174°54′E	65°56′S, 174°34′E	1830–0

TABLE 1. (continued)

Station	Date	Local Time	Position Start	Position Finish	Collecting Depth, m
colspan continued					

Station	Date	Local Time	Start	Finish	Collecting Depth, m
			Eltanin Cruise 32 (continued)		
2017	Jan. 14, 1968	1516–1603	73°59′S, 176°18′E	74°01′S, 176°26′E	239–0
2030	Jan. 17, 1968	1330–1452	75°29′S, 176°38′E	74°52′S, 176°29′E	239–0
2037	Jan. 18–19, 1968	2356–0117	75°02′S, 168°34′E	75°02′S, 168°36′E	239–0
2040	Jan. 19–20, 1968	2351–0109	76°00′S, 172°10′E	76°02′S, 172°23′E	239–0
2107	Feb. 6, 1968	0437–0546	76°59′S, 162°10′W	76°58′S, 162°25′W	460–0
2111	Feb. 9, 1968	0353–0716	74°09′S, 174°53′W	74°12′S, 174°46′W	1830–0
2133	Feb. 17, 1968	0945–1228	68°03′S, 176°05′E	68°02′S, 176°01′E	1829–0
2139	Feb. 19, 1968	1423–1712	63°59′S, 176°39′E	63°54′S, 176°32′E	1830–0
			Eltanin Cruise 33		
2174	April 21, 1968	0625–0934	63°01′S, 120°01′W	63°08′S, 120°10′W	1830–0
			Eltanin Cruise 35		
2250	Aug. 15, 1968	1705–2000	45°53′S, 132°33′E	46°05′S, 132°30′E	1150–0
2260	Aug. 22, 1968	1616–1905	56°40′S, 129°17.5′E	**56°48.2′S, 129°26.9′E**	**1200–0**
2264	Aug. 27, 1968	0637–0955	54°52′S, 128°03′E	54°47′S, 127°46.5′E	1200–0
2279	Sept. 10, 1968	0017–0250	34°49′S, 123°24′E	34°53′S, 123°05′E	1200–0
2285	Sept. 16, 1968	0729–1021	45°53′S, 117°03.3′E	46°01′S, 117°00′E	1250–0
2289	Sept. 18, 1968	1801–2010	**48°53.5′S, 117°00.1′E**	49°04′S, 117°05′E	1200–0
2293	Sept. 22, 1968	1812–2042	52°58′S, 117°08′E	53°05.2′S, 117°15′E	1300–0
2301	Sept. 29–30, 1968	2326–0200	51°03′S, 124°53′E	**50°55.2′S, 125°11.1′E**	**900–0**
			Eltanin Cruise 46		
2	Nov. 25, 1970	0227–0255	38°25′S, 115°16′E		500
	Nov. 25, 1970	0444–0517	38°26′S, 115°19′E		1000
	Nov. 25, 1970	0802–0853	38°37′S, 115°19′E		2000
4	Dec. 1, 1970	0025–0052	50°16′S, 115°42′E		100–0
	Nov. 30, 1970	2252–2350	50°16′S, 115°43′E	50°16′S, 115°41′E	500–0
	Dec. 1, 1970	0257–0337	50°17′S, 115°43′E	50°17′S, 115°39′E	1000
5	Dec. 4, 1970	0158–0232	54°04′S, 115°01′E	54°05′S, 115°02′E	500
5	Dec. 4, 1970	0415–0505	54°07′S, 115°03′E	54°12′S, 115°05′E	1000
6	Dec. 7, 1970	1816–1846	59°14′S, 114°59′E	59°15′S, 114°58′E	100
	Dec. 7, 1970	1928–2003	59°16′S, 114°58′E	59°18′S, 114°57′E	500
	Dec. 7, 1970	1630–1745	59°11′S, 115°00′E	59°12′S, 114°59′E	1000–0
7	Dec. 10, 1970	0258–0330	62°18′S, 114°38′E		100
	Dec. 10, 1970	0419–0459	62°19′S, 114°36′E	62°22′S, 114°33′E	500
8	Dec. 11, 1970	0834–0917	64°22′S, 114°27′E	64°23′S, 114°19′E	100–0
	Dec. 11, 1970	0932–1047	64°23′S, 114°20′E	64°22′S, 114°20′E	500–0
9	Dec. 14, 1970	1416–1448	64°25′S, 105°04′S	64°24′S, 105°00′E	100–0
	Dec. 14, 1970	1105–1135	64°29′S, 105°02′E		500
	Dec. 14, 1970	1246–1326	64°27′S, 105°04′E	64°25′S, 105°04′E	1000
	Dec. 14, 1970	1502–1722	64°24′S, 105°00′E	64°24′S, 104°37′E	1000–0
10	Dec. 17, 1970	2155–2240	63°44′S, 94°54′E	63°43′S, 94°54′E	100–0
	Dec. 17–18, 1970	2249–0027	63°43′S, 94°54′E	63°38′S, 94°55′E	500–0
	Dec. 18, 1970	0036–0255	63°38′S, 94°55′E	63°31′S, 95°00′E	1000–0
11	Dec. 21, 1970	0656–0726	63°17′S, 90°16′E	63°18′S, 90°13′E	100
	Dec. 21, 1970	0817–0852	**63°20′S, 90°11′E**	63°21′S, 90°08′E	500
	Dec. 21, 1970	1012–1052	63°23′S, 90°02′E	63°25′S, 89°56′E	1000
15	Jan. 1, 1971	1548–1623	63°10′S, 75°24′E	63°10′S, 75°17′E	500
	Jan. 1, 1971	1736–1806	63°08′S, 75°13′E		1000
16	Jan. 4, 1971	2226–2303	56°07′S, 73°44′E	56°05′S, 73°41′E	100–0
	Jan. 4–5, 1971	2347–0017	56°02′S, 73°39′E	56°02′S, 73°37′E	500
	Jan. 5, 1971	0127–0157	56°01′S, 73°36′E		1000
17	Jan. 9, 1971	1507–1542	50°13′S, 74°28′E	50°15′S, 74°26′E	100–0
	Jan. 9, 1971	1623–1653	50°16′S, 74°25′E		500
	Jan. 9, 1971	1751–1801	50°18′S, 74°24′E	50°19′S, 74°25′E	731
			Atlantis II Cruise 31		
RHB 1434	March 11, 1967	1004–1405	26°30′S, 36°33′W	26°43′S, 36°49′W	1875–0
RHB 1440	March 17, 1967	1118–1449	34°43′S, 49°28′W	34°51′S, 49°44′W	1295–0
RHB 1441	March 18–19, 1967	2342–0113	36°45′S, 53°06′W	36°39′S, 53°13′W	190–0

found to be significantly different from *Scolecithrix danae*, the type species of the genus. Sars [1902] also considered some of the *Scolecithrix* species that Giesbrecht [1888, 1892] described to be referable to *Scolecithricella*, namely, *S. vittata*, *S. tenuiserrata*, *S. profunda*, *S. longipes*, *S. abyssalis*, *S. dubia*, *S. dentata*, *S. marginata*, and *S. longifurca*. However, *S. longifurca* was subsequently found [Farran, 1929] to be a *Scaphocalanus*. The number of species referred to *Scolecithricella* continued to grow as many new species were described and additional species were transferred into the genus from other genera.

Sars [1925] recognized that some of the species added subsequently to *Scolecithricella* were clearly distinct from the original species in such characters as the structure of the maxillary sensory filaments and the fifth pair of legs, and he established a new genus, *Amallothrix*, to accommodate them as a separate taxon. Included originally in this new genus were *Scolecithricella gracilis* Sars, *S. propinqua* Sars, *S. curticauda* A. Scott, *S. lobata* Sars, *S. arcuata* Sars, *S. valida* (Farran), *S. emarginata* (Farran), and *S. obtusifrons* (Sars).

As was noted by Vervoort [1951], the maxillary sensory filaments seem to be basically similar in all species without showing a consistent difference between the two genera. The fifth pair of legs therefore remains the only character useful in distinguishing the species originally placed in *Amallothrix* by Sars [1925]. Vervoort [1951] further noted that certain species referred afterward to either *Scolecithricella* or *Amallothrix* showed fifth pairs of legs that were more or less intermediate, for example, *Scolecithricella fowleri* (Farran), *S. auropecten* (Giesbrecht), *Amallothrix laminata* Farran, *A. farrani* Rose, and *A. sarsi* Rose.

Bradford [1973] reviewed the family Scolecithricidae and redefined some of its genera, including *Scolecithricella* and *Amallothrix*. There are a number of species previously referred to either *Scolecithricella* or *Amallothrix* that were excluded from her redefinition of these genera, for example, *Amallophora altera* Farran, *Scolecithrix laminata* Farran, *S. ctenopus* Giesbrecht, and *S. fowleri* Farran. There are also many species that Bradford [1973] was unable to place definitely into her redefined genera because their pertinent characters had not been adequately described. The definitive classification of this complex group does not seem to be possible until most of the species are reexamined in detail. Therefore I follow Vervoort [1951] in using the generic name *Scolecithricella* for all species formerly referred to *Scolecithricella* and *Amallothrix*.

So far, the following five species of *Scolecithricella* have been originally described from antarctic waters:

S. glacialis (Giesbrecht, 1902)
S. polaris (Wolfenden, 1911)
S. incisa (Farran, 1929)
S. altera (Farran, 1929)
S. dentipes Vervoort, 1951

The following three warm-water species of *Scolecithricella* have been known to range as far south as the Antarctic:

S. valida (Farran, 1908), recorded by Farran [1929] and Vervoort [1957]
S. ovata (Farran, 1905), recorded by Farran [1929] and Vervoort [1951, 1957]
S. robusta (T. Scott, 1894), recorded by Farran [1929] and Vervoort [1957]

Of the five originally antarctic species, *S. dentipes* has been known to extend into subantarctic waters [Vervoort, 1957], and *S. glacialis* into subantarctic waters [Farran, 1929] and as far north as 42°26'S off New Zealand [Bradford, 1972], and *S. altera* has been recently found in the North Atlantic [Roe, 1975].

There is only one species of *Scolecithricella* that was originally described from subantarctic waters.

S. minor (Brady, 1883), the type species of the genus

Warm-water species that have been known to range as far south as or close to subantarctic waters are as follows:

S. dentata (Giesbrecht, 1892), recorded by Farran [1929] and Bradford [1972]
S. vittata (Giesbrecht, 1892), recorded by Far. [1929]

During the present study, all of these species which have been known to occur in antarctic and subantarctic waters have been found except for *S. incisa*. This species was originally described by Farran [1929] from a single female specimen obtained under the ice off Cape Evans in McMurdo Sound and has not been recaptured. Upon detailed morphological examination, the antarctic species *S. glacialis* (Giesbrecht, 1902) and *S. polaris* (Wolfenden, 1911) were found to be morphologically identical with *S. minor* (Brady, 1883) and *S. emarginata* (Farran,

1905), respectively. *Scolecithricella robusta*, recorded from antarctic waters by Farran [1929] and Vervoort [1957], was found to be referable to *S. dentipes* Vervoort, 1951, on the basis of intraspecific variations shown by the latter. Also found in the present study were two new records (*S. obtusifrons* and *S. profunda*) and six new species (*S. cenotelis, S. schizosoma, S. pseudopropinqua, S. parafalcifer, S. hadrosoma,* and *S. vervoorti*).

The males were extremely rare, constituting only 573 of the 5284 specimens found in the study, and males were not found in 9 of the 16 species identified in the study. However, three species of males found in the study were not specifically identified, although in all probability they belong to known species.

Key to Species of *Scolecithricella* in This Study

Female

1. No outer spine on first exopodal segment of first leg (Figure 2k)..2
   An outer spine on first exopodal segment of first leg (Figure 10k)..6
2. Viewed laterally, metasome emarginate on posterior edge (Figure 7c) 5
   Metasome not emarginate (Figure 2d)..................3
3. Rostrum with short sensory filament (Figure 2e); first inner lobe of maxillule with 1 posterior seta (Figure 2h) **S. minor**
   Rostrum with long sensory filament (Figure 4d); first inner lobe of maxillule with 2 posterior setae (Figure 4g) 4
4. Spermathecal vesicle large (Figure 4c); in fifth leg, inner spine longer than distal spine (Figure 4n) ... **S. profunda**
   Spermathecal vesicle very small (Figure 5d); in fifth leg, inner spine much shorter than distal spine (Figure 5o) .... **S. vittata**
5. First to fourth leg endopods with distal spiniform process along external margin (Figures 7k–7n); fifth leg wide (Figure 7o)............................. **S. dentata**
   First to fourth leg endopods without distal spiniform process (Figures 8k–8m); fifth leg elongate (Figure 8o) .... **S. schizosoma**, n. sp.
6. In first leg, basis without inner seta, endopod without external lobe (Figure 22k)............................15
   In first leg, basis with well-developed inner seta, endopod with external lobe armed with spinules (Figure 10k) ...7
7. Rostrum with long rami bearing short sensory filaments (Figure 10e) 8
   Rostrum with short rami bearing long sensory filaments (Figure 17d) 11
8. Spermathecal vesicle with round distal sac separated by narrow neck (Figure 15d) .... **S. valida**
   Spermathecal vesicle without such a separate distal sac (Figure 10d) 9
9. Spermathecal vesicle broad; genital segment with a row of spinules in front of genital field (Figure 13c)..........
   .... **S. pseudopropinqua**, n. sp.

Spermathecal vesicle elongate; genital segment without such a row of spinules (Figure 10d)......................10
10. Spermathecal vesicle very narrow (Figure 10d); fifth leg usually armed with spinules (Figure 10o) .... **S. dentipes**
    Spermathecal vesicle relatively wide (Figure 12c); fifth leg usually not armed with spinules (Figure 12m) .... **S. parafalcifer**, n. sp.
11. Distal segment of fifth leg pear-shaped with distal end expanded into an oval (Figure 18n); first exopodal segment of antenna without a conical process on internal margin (Figure 18f)..12
    Distal segment of fifth leg not expanded distally into an oval (Figure 19o); first exopodal segment of antenna with a conical process on internal margin (Figure 19f) 13
12. Spermathecal vesicle with distinct distal end appearing like a vacuole (Figure 18d)............... **S. cenotelis**, n. sp.
    Spermathecal vesicle without such a distal end (Figure 17c) .... **S. ovata**
13. Fifth lobe of maxilla with 2 sensory filaments (Figure 21h); mandibular basis with 3 setae (Figure 21f) **S. obtusifrons**
    Fifth lobe of maxilla with 1 sensory filament (Figure 19i); mandibular basis with 2 setae (Figure 19g)...........14
14. Lateral skeletal plate of genital orifice elongate; viewed laterally, genital field not conspicuously projected, with posterior margin nearly straight (Figure 20c); body longer than 5.00 mm .... **S. hadrosoma**, n. sp.
    Lateral skeletal plate of genital orifice not elongate; viewed laterally, genital field conspicuously projected, with posterior margin strongly curved (Figure 19c); body shorter than 5.00 mm.................. **S. emarginata**
15. Second and third inner lobes of maxillule with 2 and 4 setae, respectively (Figure 22h); posterior end of metasome in lateral view produced at an angle, spermathecal vesicle with large expansion on posterior side (Figure 22c).....
    .... **S. altera**

Second and third inner lobes of maxillule with 1 and 3 setae, respectively (Figure 23i); posterior end of metasome in lateral view roundly produced, spermathecal vesicle not expanded on posterior side (Figure 23c) **S. vervooti**, n. sp.

Male

1. No outer spine on first exopodal segment of first leg 2
   An outer spine on first exopodal segment of first leg 4
2. In fifth pair of legs, right leg exopod reaching about distal end of left leg endopod; right leg endopod moderately developed (Figure 9j) .... **S. schizosoma**, n. sp.
   In fifth pair of legs, right leg exopod extending far beyond distal end of left leg endopod; right leg endopod very small (Figure 3n) 3
3. In left fifth leg, third exopodal segment tapering into a long spiniform process (Figure 3n) .... **S. minor**
   In left fifth leg, third exopodal segment short, not tapering into a spiniform process (Figure 6h) .... **S. vittata**
4. In first leg, basis without inner seta, endopod without external lobe (Figure 23l) .... **S. vervoorti**, n. sp.
   In first leg, basis with inner seta, endopod with external lobe armed with spinules (Figure 11j) 5
5. Third inner lobe and basis of maxillule with setae greatly reduced (Figure 27d); maxillar lobes with weakly developed setae (Figures 27e and 27f) .... **S. sp. 3**
   Third inner lobe and basis of maxillule with setae relatively

well developed (Figure 11*g*); maxillar lobes with strong
setae (Figure 11*h*) 6
6. In right fifth leg, endopod extending beyond distal conical
process of first exopodal segment; second exopodal segment of right leg extending beyond distal end of left leg
basis (Figure 11*n*).................................. 7
   In right fifth leg, endopod short of reaching distal process of first exopodal segment; second exopodal segment of
   right leg reaching about distal end of left leg basis (Figure
   25*n*)... 9
7. Third exopodal segment of right fifth leg with a long distal
   lobe (Figures 16*f* and 16*g*) .... **S. valida**
   Third exopodal segment of right fifth leg without such a distal
   lobe (Figure 11*n*) 8
8. Three distal setae of maxilliped coxa poorly developed
   (Figure 11*i*)........................... **S. dentipes**
   Three distal setae of maxilliped coxa well developed (Figure
   14*d*) .... **S. pseudopropinqua**, n. sp.
9. Right fifth leg exopod terminated with a long lancet-form
   spine, as long as or longer than third segment; left fifth
   leg endopod with short distal spine (Figures 25*n* and 25*o*)
   .... **S.** sp. 1
   Right fifth leg exopod terminated with a short lancet-form
   spine, much shorter than third segment; left fifth leg
   endopod with long distal spine (Figures 26*g*–26*i*)
   .... **S.** sp. 2

Scolecithricella minor (Brady, 1883)
Figs. 2 and 3

Scolecithrix minor Brady, 1883, p. 58, pl. 16, figs. 15,
   16, pl. 18, figs. 1–5.
Scolecithricella minor; Sars, 1902, p. 55, pls. 37,
   38.—Mori, 1937, p. 51, pl. 25, figs. 1–7.—Brodsky,
   1950, p. 268, fig. 178.—Vervoort, 1965, p.
   82.—Park, 1968, p. 554, pl. 8, figs. 9–12.
Scolecithrix glacialis Giesbrecht, 1902, p. 25, pl. 4,
   figs. 1–7.
Scolecithricella glacialis; Vervoort, 1957, p. 101.
For more complete lists of synonyms of *S. minor* and
   S. glacialis, see Vervoort [1965] and Vervoort
   [1957], respectively.

Occurrence. The following station list shows the
occurrence of *S. minor* (Brady, 1883):

Eltanin Cruise 17

Sta. 17, 219–0 m, 1F (1.38 mm)
Sta. 18, 768–0 m, 2F (1.28–1.34 mm)
Sta. 20, 768–0 m, 5F (1.32–1.36 mm)
Sta. 26, 2560–0 m, 4F (1.38–1.40 mm)
Sta. 47, 457–0 m, 2F
Sta. 59, 1251–0 m, 2F (1.34–1.40 mm)
Sta. 66, 311–0 m, 2F (1.38–1.42 mm)
Sta. 69, 3146–0 m, 1F
Sta. 79, 386–0 m, 2F

Sta. 80, 625–0 m, 8F (1.26–1.40 mm)
Sta. 82, 313–0 m, 2F (1.26–1.32 mm)
Sta. 85, 625–0 m, 13F (1.26–1.32 mm)

Eltanin Cruise 21

Sta. 252, 1050–0 m, 1F (1.22 mm)
Sta. 257, 1000–0 m, 6F (1.20–1.32 mm)
Sta. 270, 128–0 m, 2F (1.28–1.30 mm)
Sta. 272, 1000–0 m, 8F (1.34 mm)

Eltanin Cruise 23

Sta. 1697, 2274–0 m, 3F (1.18 mm)
Sta. 1700, 1275–0 m, 4F (1.08–1.10 mm)
Sta. 1704, 800–0 m, 2F (1.16 mm)
Sta. 1710, 900–0 m, 2F (1.22 mm)

Eltanin Cruise 26

Sta. 1835, 1375–0 m, 8F (1.24–1.28 mm)
Sta. 1839, 3750–0 m, 6F (1.30 mm)
Sta. 1842, 1350–0 m, 6F (1.32 mm)

Eltanin Cruise 35

Sta. 2260, 1200–0 m, 1F (1.26 mm)
Sta. 2264, 1200–0 m, 12F (1.22–1.40 mm)
Sta. 2279, 1200–0 m, 1F (1.10 mm)
Sta. 2285, 1250–0 m, 2F (1.10 mm)
Sta. 2289, 1200–0 m, 6F (1.20–1.26 mm)
Sta. 2293, 1300–0 m, 16F (1.22–2.38 mm)
Sta. 2301, 900–0 m, 9F (1.20–1.32 mm)

Eltanin Cruise 46

Sta. 4,   100 m, 33F (1.28–1.46 mm)
           500 m, 15F (1.30–1.38 mm);   2M
           1000 m, 162F (1.30–1.40 mm);   22M
Sta. 5,   500 m, 1F;   2M
           1000 m, 5F (1.34–1.38 mm);   2M
Sta. 6,   100 m, 36F (1.24–1.38 mm); 5M
           500 m, 4F (1.32–1.36 mm)
           1000 m, 4F (1.32–1.38 mm);   1M
Sta. 7,   100 m, 10F (1.32–1.38 mm)
           500 m, 6F (1.34–1.38 mm)
Sta. 8,   100 m, 6F (1.32–1.38 mm)
           500–0 m, 245F (1.28–1.42 mm);
                   47M (1.34–1.46 mm)
Sta. 9,   1000–0 m, 108F (1.30–1.40 mm);   11M
           1000 m, 27F (1.28–1.36 mm); 3M
Sta. 10,   100–0 m, 11F (1.28–1.40 mm); 2M
           500–0 m, 188F (1.28–1.40 mm); 15M
           1000–0 m, 243F (1.32–1.40 mm); 35M
Sta. 11,   100 m, 2F (1.30–1.36 mm)
           500 m, 2F (1.30 mm)

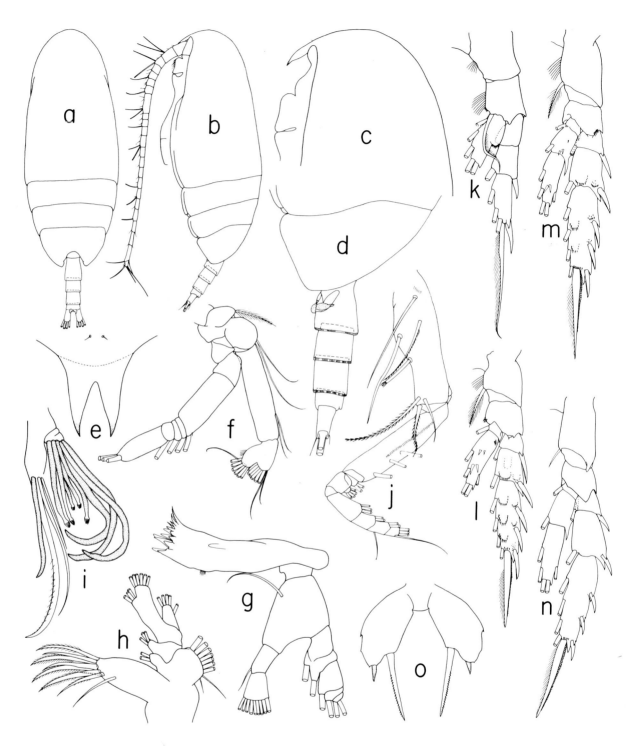

Fig. 2. *Scolecithricella minor* female: *a*, habitus, dorsal; *b*, habitus, lateral; *c*, forehead, lateral; *d*, posterior part of body, lateral; *e*, rostrum, anterior; *f*, antenna; *g*, mandible; *h*, maxillule; *i*, distal part of maxilla; *j*, maxilliped; *k*, first leg, anterior; *l*, second leg, posterior; *m*, third leg, posterior; *n*, fourth leg, posterior; and *o*, fifth pair of legs, posterior.

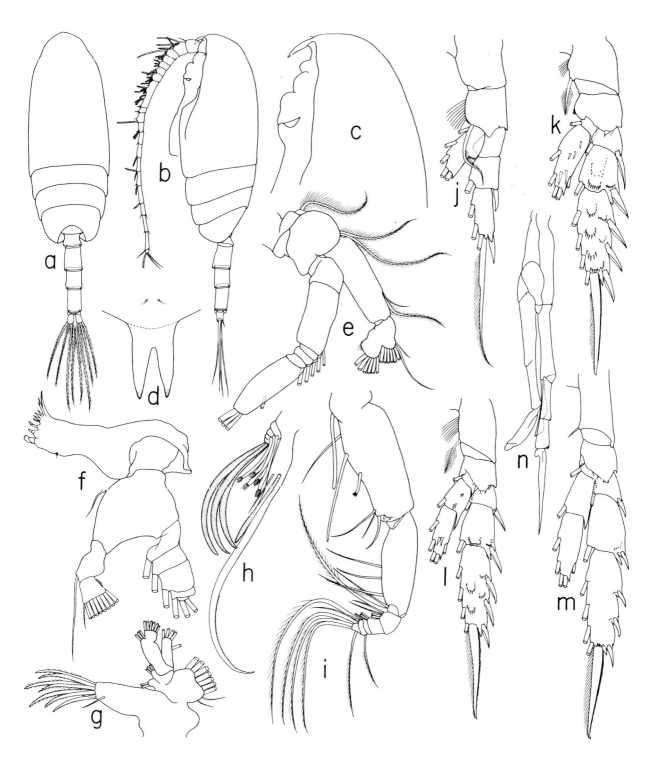

Fig. 3. *Scolecithricella minor* male: *a*, habitus, dorsal; *b*, habitus, lateral; *c*, forehead, lateral; *d*, rostrum, anterior; *e*, antenna; *f*, mandible; *g*, maxillule; *h*, distal part of maxilla; *i*, maxilliped; *j*, first leg, anterior; *k*, second leg, posterior; *l*, third leg, posterior; *m*, fourth leg, posterior; and *n*, fifth pair of legs, anterior.

1000 m, 1F (1.30 mm)
Sta. 12,   844–0 m, 103F (1.24–1.40 mm); 18M
Sta. 15,   500 m, 8F (1.32–1.34 mm); 1M
Sta. 16,   100 m, 1M
          500 m, 10F (1.28–1.38 mm); 20M
          1000 m, 1F (1.28 mm)
Sta. 17,   100 m, 86F (1.26–1.40 mm); 33M
          500 m, 28F (1.30–1.40 mm); 7M
          731 m, 11F (1.32–1.34 mm); 6M

Total: 1495F and 233M

Female. Prosome length, 0.92–1.18 mm; body length, 1.08–1.46 mm. Prosome in dorsal view narrowly elliptical (Figure 2*a*). First metasomal segment fused with cephalosome. Fourth and fifth metasomal segments fused, without a visible articulation suture. Forehead narrowly rounded in dorsal view but broadly rounded in lateral view (Figure 2*c*). Posterolateral corners of metasome symmetrical, produced distally covering anterior 1/3 of genital segment. In lateral view (Figure 2*d*), metasome attenuated posteriad ending in an obtuse angle.

Urosome about 1/4 length of prosome. In dorsal view, urosome 1/6–1/5 as wide as prosome. Urosome 4-segmented, genital segment longest, third segment slightly longer than second. Caudal rami nearly as long as third urosomal segment. Viewed laterally (Figure 2*d*), spermatheca tapered distally into a small, straight digitiform vesicle. Lateral skeletal plate of genital orifice elongated, extending obliquely forward. Rostrum biramous, strong, but distal ends of rami transformed into soft, transparent sensory filaments (Figure 2*e*).

Antennule (Figure 2*b*) with 23 free segments (because the eighth, ninth, and tenth of the 25 segments are fused), reaching about posterior end of genital segment. In antenna (Figure 2*f*), exopod longer than endopod; first exopodal segment without setae; second and third exopodal segments fused, with a minute seta belonging to second segment and a large seta to third. Seventh exopodal segment with 3 distal setae only. Mandible (Figure 2*g*) with long masticatory blade. Endopod and exopod of mandible about equal in length; basis and first endopodal segment each with a single seta. In maxillule (Figure 2*h*), first inner lobe with 1 large posterior seta, 7 large and 1 small distal setae. Second and third inner lobes with 2 and 3 setae, respectively. Basis with 3 setae, endopod with 6 setae, exopod with 5 setae, and

outer lobe with 9 setae. First 3 lobes of maxilla each with 3 setae, fourth lobe with 2 setae, fifth lobe with 3 setae and 1 vermiform sensory filament. Sixth lobe absent. Distal part (endopod) of maxilla (Figure 2*i*) with 3 long vermiform and 5 brush-form sensory filaments. Coxa of maxilliped (Figure 2*j*) with 2 vermiform and 1 brush-form sensory filaments and 1 seta on anterior half and 3 setae on posterior end. Basis with 3 middle and 2 terminal setae. Five endopodal segments, in order from proximal to distal, with 4, 3, 2, 2 + 1, and 3 + 1 setae.

In first leg (Figure 2*k*), basipod with inner marginal hair on both coxa and basis, 1 inner distal seta on basis, and some spinules on basis between inner marginal hair and inner distal seta. Endopod 1-segmented, with spinous outer lobe and 5 setae. External distal corner of endopod produced into a spiniform process. Exopod 3-segmented; first segment without setae or spines; second segment with 1 inner seta and 1 external spine; third segment with 3 inner setae, 1 terminal seta, and 1 outer spine.

In second leg (Figure 2*l*), basipod with inner marginal hair and 1 inner seta on coxa and some conspicuous inner marginal spinules on basis. Endopod 2-segmented; first segment with 1 seta; second segment with 5 setae. External distal corner of endopod produced into a spiniform process. Exopod 3-segmented; first 2 segments each with 1 inner seta and 1 outer spine; third segment with 4 inner setae, 3 outer spines, and 1 terminal spine. All outer spines of about equal length, each extending beyond proximal end of following spine. Second endopodal and second and third exopodal segments armed with spinules on posterior surface.

Basipod of third leg (Figure 2*m*) similar to that of second leg, but basis without inner marginal spinules. Endopod 3-segmented; first 2 segments each with 1 seta; third segment with 5 setae. External distal corner of endopod produced into a spiniform process. Exopod 3-segmented, with same meristic details as in second leg; but outer spines are relatively shorter. Distal 2 segments of both rami armed with spinules on posterior surface.

Fourth leg (Figure 2*n*) similar to third except that coxa without inner marginal hair, inner seta of coxa small and not plumose, exopod with outer spines relatively small, and posterior surface of rami without conspicuous spinules other than a row of spinules at base of terminal spine of exopod.

Fifth leg (Figure 2*o*) uniramous, 1-segmented, flat, and attached to common basal coupler; with a minute external spine, a small distal spine, and a large inner spine.

Male. Prosome length, 0.92–1.06 mm; body length, 1.34–1.46 mm. Body slenderer than that of female. First metasomal segment fused with cephalosome. Fourth and fifth metasomal segment fused. Urosome about $2/5$ length of prosome. Second and fourth urosomal segments about equal in length and longer than other urosomal segments. Laterally (Figure 3*b*), forehead narrower, and posterior end of metasome less produced than female. Viewed anteriorly (Figure 3*d*), rostrum similar to but slenderer than that of female.

Antennules symmetrical, with 20 free segments (Figure 3*b*). Of 25 segments, eighth through twelfth as well as twentieth and twenty-first are fused. When applied against body, antennules reach as far as posterior end of second urosomal segment. Antenna (Figure 3*e*) similar to but better developed than that of female. Seta belonging to second exopodal segment well developed. Seventh exopodal segment with a well-developed middle seta in addition to 3 terminal setae. In mandible (Figure 3*f*), masticatory blade slightly reduced as compared with that of female. Mandibular palp with same number of setae as in female, but seta on basis smaller, and seta on first endopodal segment larger than corresponding seta in female. Maxillule (Figure 3*g*) somewhat reduced in size but same in setation as in female. Maxilla (Figure 3*h*) similar to that of female but third and fourth lobes each with 2 setae. Maxilliped (Figure 3*i*) somewhat reduced in length but with better developed endopodal setae than in female. Setation of maxilliped as in female, but second and third endopodal segments with 2 and 1 setae, respectively, instead of 3 and 2.

First to fourth legs (Figures 3*j*–3*m*) similar to those of female. Fifth pair of legs (Figure 3*n*) strongly asymmetrical, with left leg much longer than right leg. Both legs biramous, but right leg with very small endopod. Right exopod 3-segmented, with first and second segments partially fused; second segment reaching about middle of first left exopodal segment. Left leg with 1-segmented endopod and 3-segmented exopod. Endopod reaching middle of second exopodal segment. Third exopodal segment tapering as a spiniform process, longer than combined length of first and second segments.

Remarks. *Scolecithricella minor* was originally described as *Scolecithrix minor* by Brady [1883] from female and male specimens obtained in a surface net tow from 46°46′S, 45°31′E in the southern

Indian Ocean. Sars [1902] recorded *Scolecithrix minor* Brady in Norwegian waters and provided a detailed redescription of the species. By recognizing differences of *Scolecithrix minor* from *Scolecithrix danae*, the type species of the genus, Sars [1902] established the genus *Scolecithricella* to accommodate the former.

Scolecithricella minor has been known to occur widely in all the great oceans [Vervoort, 1965], but most of its distribution records are from high latitudes of both the northern and the southern hemisphere.

Scolecithricella glacialis was originally described as *Scolecithrix glacialis* by Giesbrecht [1902] from female specimens collected by the *Belgica* expedition from between 69°48′S and 71°18′S in latitude and between 81°19′W and 92°22′W in longitude. It has been known to occur widely in antarctic and subantarctic waters and as far north as 44°05′S, 147°35′E [Vervoort, 1957].

According to the original descriptions it is impossible to distinguish Giesbrecht's *glacialis* from Brady's *minor*. However, Giesbrecht [1902] gave no reference to Brady's *minor*, though he attempted to compare his *glacialis* with *Scolecithricella abyssalis* and *S. dentata*, rather remotely related species.

In the present study, 1495 females and 233 males have been found that are referable to either *S. minor* (Brady, 1883) or *S. glacialis* (Giesbrecht, 1902). They were found throughout the antarctic and subantarctic seas and as far north as about 35°S, though they were more common in antarctic waters, and showed no significant morphological variations except for a south to north gradient in size. Therefore it is believed that all specimens found in the present study belong to a single species which is referable to *S. minor* (Brady, 1883) and that *S. glacialis* (Giesbrecht, 1902) is a junior synonym.

The descriptions of *S. minor* from the northern hemisphere by various authors [e.g., Sars, 1902; Mori, 1937] seem to agree with the antarctic and subantarctic specimens examined in the present study. However, Brodsky [1950] recognized two forms of *S. minor: orientalis* from the North Pacific and *occidentalis* from the North Atlantic, which he distinguished from each other mainly by the form of the fifth legs. The specimens found in the present study were similar to *occidentalis*.

In the present study, *S. minor* was found to be the most common species of the genus. About 33% of the *Scolecithricella* specimens found in the study belonged to this species. It seemed to be the only

Scolecithricella species that inhabits mainly the epipelagic parts of the antarctic seas.

Scolecithricella profunda (Giesbrecht, 1892)
Fig. 4

Scolecithrix profunda Giesbrecht, 1892, p. 266, pl. 13, figs. 5, 26.
Scolecithricella profunda; A. Scott, 1909, p. 91.—Farran, 1936, p. 97.—Tanaka, 1962, p. 45, fig. 131.—Vervoort, 1965, p. 80.

Occurrence. The following station list shows the occurrence of *S. profunda* (Giesbrecht, 1892):

***Eltanin* Cruise 21**
Sta. 213, 1050-0 m, 1F (2.20 mm)
Sta. 257, 1000-0 m, 1F (2.10 mm)

***Eltanin* Cruise 35**
Sta. 2285, 1250-0 m, 1F (2.02 mm)

***Eltanin* Cruise 46**
Sta. 2, 500 m, 26F (1.98-2.14 mm)

Total: 29F

Female. Prosome length, 1.70-1.80 mm; body length, 1.98-2.20 mm. Similar in habitus to *Scolecithricella minor*, but in lateral view (Figure 4a), head attenuated anteriad, and posterior end of metasome broadly rounded. Urosome about ¼ length of prosome and, in dorsal view, about ⅕ as wide as prosome. Of urosomal segments, genital segment longest; second segment only slightly longer than third. Caudal ramus shorter than third urosomal segment. In lateral view (Figure 4c), spermatheca extended dorsad as an elongate, digitiform vesicle, slightly curved forward. Lateral skeletal plate of genital orifice extending in a dorsoanterior direction as an elongated triangle reaching close to anterior end of segment. Rostrum (Figure 4d) with well-developed rami, each tapering into a long transparent sensory filament.

Antennule (Figure 4a) with 23 free segments, eighth through tenth segments being fused to form a long article and extending beyond distal end of caudal ramus by last 2 segments. Antenna (Figure 4e) and mandible (Figure 4f) agree in meristic details with those of *S. minor*, but second and third exopodal segments of antenna partially separate, seta on second segment relatively long, and endopod of mandible relatively short.

Maxillule (Figure 4g) better developed than in *S.*

minor, with 2 large posterior setae and 9 large and 1 small distal setae on first inner lobe, 2 setae on second, 3 setae on third; 5 setae on basis, 3 + 5 setae on endopod, 6 setae on exopod, and 9 setae on outer lobe. Maxilla (Figure 4h) as in *S. minor*. Five endopodal segments of maxilliped (Figure 4i) with, in order from anterior to posterior, 4, 3, 3, 3 + 1, and 4 setae.

Swimming legs (Figures 4j-4m) as in *S. minor* except for some minor differences as noted below. In first to fourth legs, external distal end of endopod not produced into a spiniform process. Outer exopodal spines of first to fourth legs relatively small except for one on first exopodal segment of second leg, which is long and curved. In basis of second leg, inner marginal spinules not conspicuously strong. Spinules on posterior surface of second leg fewer in number and not arranged in arcs as in *S. minor*. Posterior surface of third and fourth legs covered widely and densely by small spinules. Fifth pair of legs (Figure 4n) similar to that of *S. minor*, but inner spine less than twice as long as distal spine. In *S. minor*, inner spine about 5 times as long as distal spine. Fifth leg of *S. profunda* is distinguishable from that of *S. abyssalis* as figured by Giesbrecht [1892] by its rounded distal margin and relatively short spines.

Remarks. *Scolecithricella profunda* was originally described as *Scolecithrix profunda* by Giesbrecht [1892] from female specimens obtained in the Golfo di Napoli (Gulf of Naples) in the western part of the Mediterranean. The species has been recorded from the Malay Archipelago by A. Scott [1909], the Great Barrier Reef area by Farran [1936], the Izu region of Japan by Tanaka [1962], and the Gulf of Guinea by Vervoort [1965]. The male has been described by Tanaka [1962] from the Izu region of Japan.

Scolecithricella profunda is practically identical in body size and morphological details to *Scolecithricella abyssalis* (Giesbrecht, 1888). The latter was originally described from deep waters of the tropical Pacific and has been known to occur in almost all geographical areas where the former has been recorded [Tanaka, 1962]. In the southern hemisphere, *S. abyssalis* has been recorded as far south as 29°S in the western Indian Ocean by Wolfenden [1911]. As was suggested by Vervoort [1965], it is quite likely that these two species are the same. According to Giesbrecht [1892], however, the two species are distinguishable by small differences in the shape of the fifth leg. All specimens found in the present study had fifth legs similar to those of *S. profunda* as figured by Giesbrecht [1892].

In the present study, *S. profunda* was found in small numbers at four northern stations, all north to 49°10'S. This finding represents the first record of the species in subantarctic waters.

Scolecithricella vittata (Giesbrecht, 1892)
Figs. 5 and 6

Scolecithrix vittata Giesbrecht, 1892, p. 266, pl. 13, figs. 2, 23, 32, 34; pl. 37, figs. 5, 8.
Scolecithricella vittata; Sars, 1925, p. 190, pl. 52, figs. 15–20.—Farran, 1926, p. 259; 1929, p. 247.—Rose, 1933, p. 158, fig. 173.—Farran, 1936, p. 97.—Rose, 1942, p. 140, figs. 31–34.—Wilson, 1950, p. 335, pl. 18, figs. 233, 234,—Grice, 1962, p. 208, pl. 17, figs. 1–8.—Tanaka, 1962, p. 41, fig. 129.—Owre and Foyo, 1967, p. 61, figs. 385–388.—Park, 1968, p. 555.

Occurrence. The following station list shows the occurrence of *S. vittata* (Giesbrecht, 1892):

Eltanin Cruise 23

Sta. 1704,　800–0 m, 1F (1.82 mm)

Eltanin Cruise 35

Sta. 2279, 1200–0 m, 3F (1.76 mm)
Sta. 2285, 1250–0 m, 3F (1.78–1.80 mm)

Eltanin Cruise 46

Sta. 2,　500 m, 172F (1.60–1.78 mm);
　　　　　　42M (1.62–1.72 mm)

Total: 179F and 42M

Female. Prosome length, 1.36–1.48 mm; body length, 1.60–1.82 mm. Most similar in habitus to *Scolecithricella profunda*. Body in lateral view with attenuated head (Figure 5*b*) and broadly rounded posterior edge of metasome (Figure 5*c*). Urosome about ⅕ length of prosome. Genital segment about twice length of second urosomal segment (Figure 5*c*). Second and third urosomal segments of about equal length. In lateral view (Figure 5*d*), spermatheca extended dorsad as a small, straight digitiform vesicle, much smaller than in *S. profunda*. Lateral skeletal plate of genital orifice large and in the form of an elongated ellipse lying in an anterodorsal direction. Viewed laterally, genital field elevated along posterior edge, forming a step from

ventral surface of segment. This step is also present in *S. profunda* but is not conspicuous. Rostral rami strong, in anterior view curved inward (Figure 5*e*), each with a long, transparent sensory filament. Antennule with 23 free segments as in *S. profunda* but reaching only as far as posterior end of caudal ramus. Antenna (Figure 5*f*), mandible (Figure 5*g*), maxillule (Figure 5*h*), maxilla (Figure 5*i*), and maxilliped (Figure 5*j*) agree in all anatomical details with those of *S. profunda*.

First to fourth legs (Figures 5*k*–5*n*) also similar to those of *S. profunda*, with endopod not produced distally into a spiniform process along external edge and first exopodal spine of second leg considerably larger than second and third exopodal spines. Inner marginal spinules on second leg basis found in *S. profunda* and *S. minor* are absent. Posterior surface of second to fourth legs armed with spinules, but the spinules are neither so well arranged into groups as in *S. minor* nor so numerous as in *S. profunda*. Fifth leg (Figure 5*o*) 1-segmented, lamelliform, attached to a common basal coupler, similar in shape of segment to *S. minor* or *S. profunda*, but distinct in having a long distal spine nearly as long as segment itself. Inner spine about ½ length of distal spine.

Male. Prosome length, 1.22–1.28 mm; body length, 1.62–1.72 mm. Similar in habitus to female except that posterior end of metasome in lateral view relatively narrow (Figure 6*b*). Urosome about ⅕ length of prosome. Second and fourth urosomal segments about equal in length and longer than others.

Antennules symmetrical, with eighth to twelfth segments fused, forming a long article, and twentieth and twenty-first segments also fused as in *S. minor*. When applied against body, antennule reaches close to distal end of urosome. Antenna (Figure 6*c*) and mandible (Figure 6*d*) as in *S. minor* with exception that seta on first endopodal segment of mandible relatively short. Maxillule (Figure 6*e*) with same number of setae on various lobes as in female, but setae on inner lobes and basis are poorly developed. Maxilla (Figure 6*f*) similar to that of female, with setae on first to fifth lobes only slightly reduced in size. Maxilliped (Figure 6*g*) similar to that of *S. minor*, but 5 endopodal segments bearing 4, 3, 3, 3 + 1, and 3 + 1 setae, in order from proximal to distal, instead of 4, 2, 1, 2 + 1, and 3 + 1.

First to fourth legs similar to those of female except that outer spine of first exopodal segment in second and third legs relatively long. Fifth pair of legs, when applied against body, extends beyond dis-

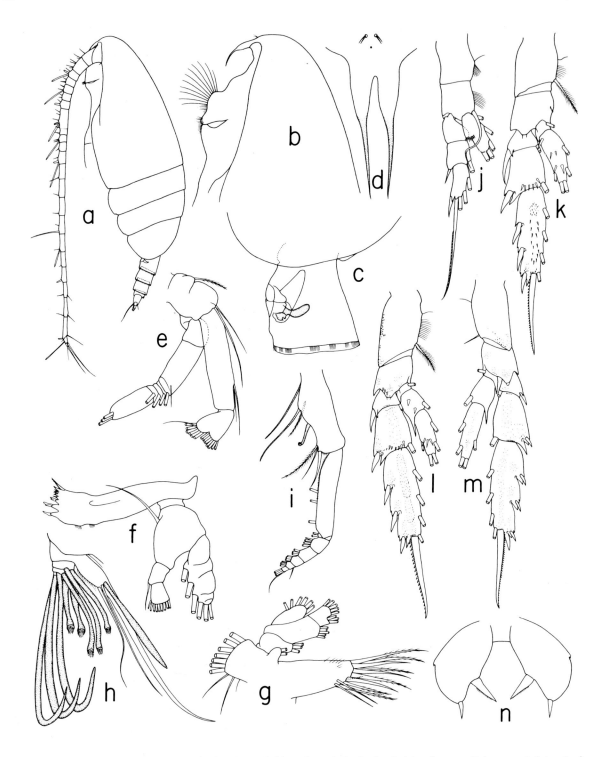

Fig. 4. *Scolecithricella profunda* female: *a*, habitus, lateral; *b*, forehead, lateral; *c*, genital segment, lateral; *d*, rostrum, anterior; *e*, antenna; *f*, mandible; *g*, maxillule; *h*, distal part of maxilla; *i*, maxilliped; *j*, first leg, anterior; *k*, second leg, posterior; *l*, third leg, posterior; *m*, fourth leg, posterior; and *n*, fifth pair of legs, posterior.

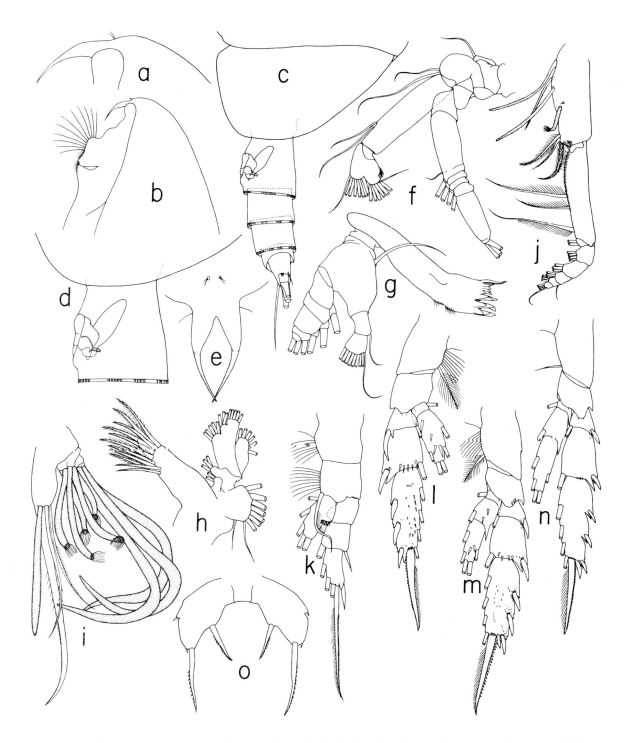

Fig. 5. *Scolecithricella vittata* female: *a*, anterior end of head, lateral; *b*, forehead, lateral; *c*, posterior part of body, lateral; *d*, genital segment, lateral; *e*, rostrum, anterior; *f*, antenna; *g*, mandible; *h*, maxillule; *i*, distal part of maxilla; *j*, maxilliped; *k*, first leg, anterior; *l*, second leg, posterior; *m*, third leg, posterior; *n*, fourth leg, posterior; and *o*, fifth pair of legs, posterior.

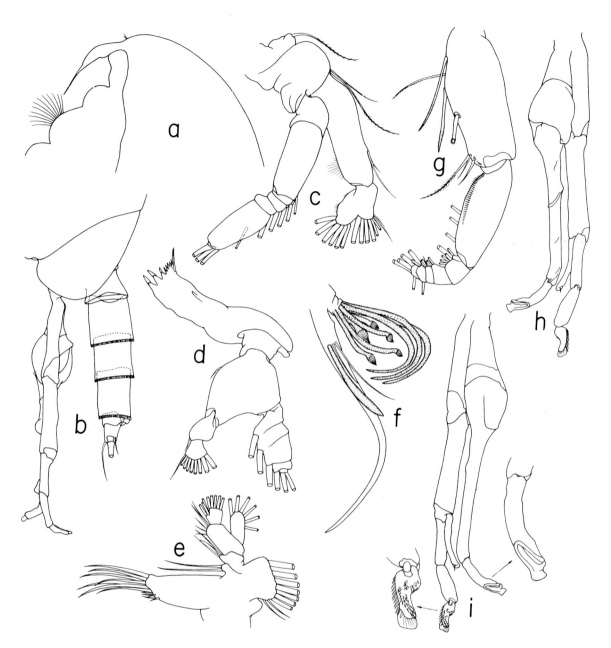

Fig. 6. *Scolecithricella vittata* male: *a*, forehead, lateral; *b*, posterior part of body, lateral; *c*, antenna; *d*, mandible; *e*, maxillule; *F*, distal part of maxilla; *g*, maxilliped; *h*, fifth pair of legs, anterior; and *i*, fifth pair of legs, posterior.

tal end of urosome by about ⅓ its length (Figure 6*b*). In right leg (Figures 6*h*–6*i*), endopod minute; exopod 3-segmented, with second segment reaching close to posterior end of first exopodal segment of left leg. Third exopodal segment terminating with a characteristic grooved structure. Left leg with 1-segmented endopod and 3-segmented exopod. Endopod about equal in length to first exopodal segment, with a small seta distally. Third exopodal segment short-

est, heavily armed with spinules on posterior surface.

Remarks. *Scolecithricella vittata* was originally described as *Scolecithrix vittata* by Giesbrecht [1892] from female specimens 1.65 mm long collected in the Golfo di Napoli. The species has been recorded from the temperate North Atlantic [Sars, 1925], the Bay of Biscay [Farran, 1926], the Florida Current [Owre and Foyo, 1967], off New Zealand [Farran, 1929], the Great Barrier Reef [Farran, 1936], the

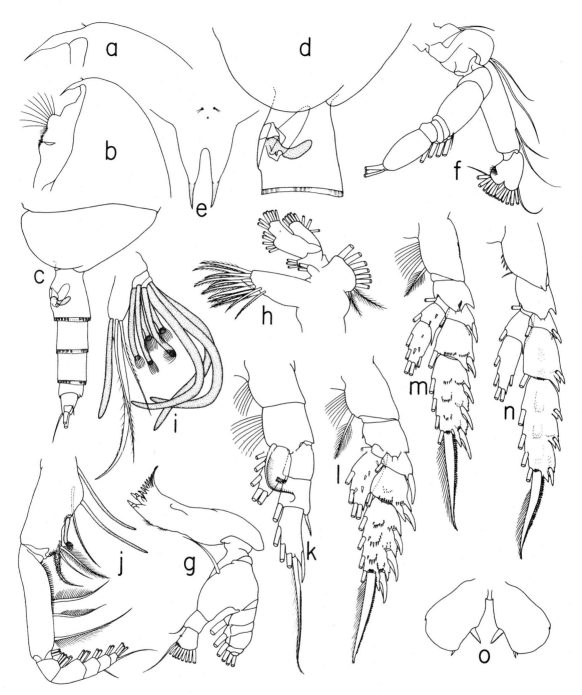

Fig. 7. *Scolecithricella dentata* female: *a*, anterior end of head, lateral; *b*, forehead, lateral; *c* posterior part of body; *d*, genital segment, lateral; *e*, rostrum, anterior; *f*, antenna; *g*, mandible; *h*, maxillule; *i*, distal part of maxilla; *j*, maxilliped; *k*, first leg, anterior; *l*, second leg, posterior; *m*, third leg, posterior; *n*, fourth leg, posterior; and *o*, fifth pair of legs, posterior.

equatorial and temperate Pacific [Wilson, 1950; Grice, 1962; Park, 1968], and off the Pacific coast of middle Japan [Tanaka, 1962].

The male has been briefly described by Tanaka [1962] from a specimen 1.62 mm long taken from deep waters off the Pacific coast of middle Japan. In the present study, 179 females and 42 males were found at four stations north to 45°53′S off Australia and New Zealand. All specimens seemed to agree with descriptions by Giesbrecht [1892] and Tanaka [1962].

Scolecithricella dentata (Giesbrecht, 1892)
Fig. 7

Scolecithrix dentata Giesbrecht, 1892, p. 266, pl. 13, figs. 12, 20, 33, pl. 37, figs. 13, 14.—Farran, 1905, p. 35.

Scolecithricella dentata; Farran, 1908, p. 51.—Sars, 1925, p. 191, pl. 52, figs. 21-23.—Farran, 1926, p. 259; 1929, p. 247; 1936, p. 97.—Rose, 1942, p. 144, 154, figs. 38-41, 46-48.—Wilson, 1950, p. 333, pl. 18, fig. 231 (female only).—Tanaka, 1962, p. 42, fig. 130.—Owre and Foyo, 1967, p. 61, figs. 94, 379-381.—Park, 1968, p. 555, pl. 18, figs. 13-16.

Scolecithrix dubia Giesbrecht, 1892, p. 266, pl. 13, fig. 29.

Occurrence. The following station list shows the occurrence of *S. dentata* (Giesbrecht, 1892):

Eltanin Cruise 21

Sta. 212, 500-0 m, 1F (1.32 mm)
Sta. 240, 2470-0 m, 2F (1.52-1.62 mm)

Eltanin Cruise 23

Sta. 1697, 2274-0 m, 2F (1.40-1.42 mm)
Sta. 1704, 800-0 m, 1F (1.40 mm)
Sta. 1710, 900-0 m, 1F (1.48 mm)

Eltanin Cruise 26

Sta. 1825, 1625-0 m, 4F (1.50-1.52 mm)
Sta. 1835, 1375-0 m, 1F

Eltanin Cruise 35

Sta. 2279, 1200-0 m, 1F
Sta. 2285, 1250-0 m, 1F (1.54 mm)

Eltanin Cruise 46

Sta. 2, 500 m, 64F (1.34-1.56 mm)

Total: 78F

Female. Prosome length, 1.06-1.22 mm; body length, 1.32-1.62 mm. Similar in habitus to *Scolecithricella vittata* and *S. profunda*, but distal margin of metasome with a large incision and urosome relatively long (about 1/3.5 length of prosome). Viewed laterally (Figure 7b), head slightly attenu- ated anteriad. Rostrum (Figure 7e) with strong rami, each tapering distally into a short, transparent sensory filament.

Genital segment nearly 1.5 times as long as second or third urosomal segment (Figure 7c). Viewed laterally, genital segment (Figure 7d) most similar to that of *S. profunda*, but spermatheca and lateral skeletal plate of genital orifice considerably larger and posterior step of genital field far from distal end of segment.

Antennule reaching posterior end of second urosomal segment, with eighth to tenth segments fused, thereby reducing number of free segments to 23. Meristic details of antenna (Figure 7f) and man- dible (Figure 7g) agree with those of *S. profunda*. Maxillule (Figure 7h) with 2 + 5 setae on endopod and 8 setae on exopod. Setation of other lobes as in *S. profunda*. Maxilla (Figure 7i) also as in *S. profunda*. Maxilliped endopod (Figure 7j) with 4, 3, 2, 2 + 1, and 4 setae on segments 1-5.

Endopod in first to fourth legs (Figures 7k-7n) produced distally into a spiniform process along external margin. Coxa of fourth leg with a row of conspicuous spinules along inner margin. Basis of second leg with a row of short but strong spinules on inner margin. Second leg exopod with outer spines of about similar size. Outer spines on third leg exopod gradually increase in size toward distal end of leg. Fourth leg exopod with outer spines of equally small size. Exopods of second and third legs with relatively large spinules arranged in arcs as in *S. minor*. In fourth leg, second and third exopodal segments armed with fine spinules on posterior surface. Fifth legs (Figure 7o) 1-segmented, flat, squarish, attached to common basal coupler, with a minute external spine, 1 or 2 small distal spines, and a relatively large inner spine.

Remarks. *Scolecithricella dentata* was originally described as *Scolecithrix dentata* by Giesbrecht [1892] from female specimens 1.3-1.45 mm long obtained in the Golfo di Napoli. The descriptions were accompanied by figures of the body in lateral view, second leg, fourth leg basipod, and fifth leg. Rose [1942] redescribed the species with a complete illustration of both the female and the male from the Baie d'Alger (Bay of Algiers) in the western Mediterranean Sea, which is close to the type locality.

Giesbrecht [1892] described a male under the name of *Scolecithrix dubia* from the Golfo di Napoli. The only figure given in the description was the fifth pair of legs, which agrees well with that of an *S. dentata* male as figured by Rose [1942]. Therefore Giesbrecht's [1892] *S. dubia* is here considered synonymous with *S. dentata*.

Scolecithricella dentata has been recorded from the Mediterranean Sea [Giesbrecht, 1892; Rose, 1942], the North Atlantic [Farran, 1905, 1908, 1926; Sars, 1925; Wilson, 1950], the Florida Current [Owre and Foyo, 1967], the North Pacific [Wilson, 1950; Tanaka, 1962; Park, 1968], the equatorial Pacific [Wilson, 1950], and the South Pacific [Farran, 1929, 1936].

Scolecithricella dentata is characteristic in having a large incision on the distal margin of the prosome and squarish fifth legs. The present specimens agree well with descriptions and figures given by Giesbrecht [1892] and Rose [1942], including details of the leg endopods, which have a spiniform process distally along the external margin. Tanaka [1962], however, figured leg endopods that show no spiniform process on the distal end.

In the present study a total of 78 females were found from 10 northern stations, all north to 46°40'S.

Scolecithricella schizosoma, n. sp.
Figs. 8 and 9

Occurrence. The following station list shows the occurrence of *S. schizosoma*, n. sp.:

Eltanin Cruise 17

Sta. 26, 2560–0 m, 1F (2.08 mm)
Sta. 62, 1251–0 m, 1F
Sta. 63, 1251–0 m, 3F (2.10 mm)
Sta. 77, 836–0 m, 1F
Sta. 80, 625–0 m, 1F (2.08 mm); 1M (2.10 mm)
Sta. 85, 625–0 m, 1F (1.76 mm)
Sta. 88, 2502–0 m, 1F (1.82 mm)

Eltanin Cruise 21

Sta. 264, 1230–0 m, 1F (1.98 mm)
Sta. 272, 1000–0 m, 1F

Eltanin Cruise 23

Sta. 1704, 800–0 m, 7F (1.74–1.78 mm)
Sta. 1710, 900–0 m, 5F (1.80 mm)

Eltanin Cruise 26

Sta. 1825, 1625–0 m, 7F (1.80–1.82 mm)
Sta. 1835, 1375–0 m, 1F (1.66 mm)
Sta. 1836, 2181–0 m, 1F (2.02 mm)
Sta. 1839, 3750–0 m, 1F

Eltanin Cruise 35

Sta. 2301, 900–0 m, 1F (1.80 mm)

Eltanin Cruise 46

Sta. 2,   500 m, 41F (1.70–1.98 mm)
Sta. 4,   500 m, 1M
           1000 m, 4F (2.04–2.06 mm)
Sta. 5,   1000 m, 3F (1.94–2.06 mm)
Sta. 6,   1000 m, 1M (2.16 mm)
Sta. 7,   500 m, 7F (2.00–2,10 mm);
           1M (2.28 mm)
Sta. 8,   500–0 m, 2F (1.92–2.00 mm);
           1M (2.20 mm)
Sta. 9,   1000–0 m, 17F (1.94–2.16 mm);
           4M (2.20–2.28 mm)
           1000 m, 4F (1.98–2.06 mm)
Sta. 10,   500–0 m, 1F (1.96 mm)
           1000–0 m, 22F (1.94–2.08 mm):
           2M (2.18–2.20 mm)
Sta. 11, 1000 m, 2F (1.94 mm)
Sta. 16, 1000 m, 3F (2.00–2.02 mm)

Total: 140F and 11M

Female. Prosome length, 1.38–1.66 mm; body length, 1.66–2.16 mm. Very close morphologically to *Scolecithricella dentata* but, as described below, shows enough differences from it to warrant the separate species status.

Laterally (Figure 8a), prosome relatively broad. Urosome about ¼ length of prosome. Incision on distal edge of metasome rerelatively shallow. Distal vesicle of spermatheca (Figure 8c) somewhat wider, and lateral skeletal plate of genital orifice less attenuated than in *S. dentata*. Antennule with eighth through tenth segments fused, reaching posterior end of second urosomal segment. Seventh exopodal segment of antenna (Figure 8f) with a small middle seta in addition to 3 distal setae. Mandible (Figure 8g), maxillule (Figure 8h) maxilla (Figure 8i), and maxilliped (Figure 8j) agree in all meristic details with those of *S. dentata*. Endopod in first to fourth legs (Figures 8k–8n) not produced distally into a

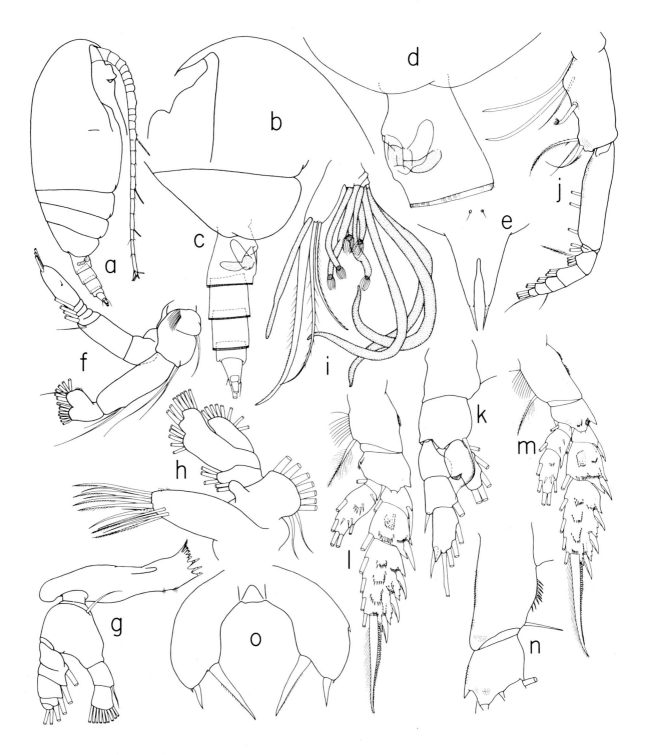

Fig. 8. *Scolecithricella schizosoma*, n. sp., female: *a*, habitus, lateral; *b*, forehead, lateral; *c*, posterior part of body, lateral; *d*, genital segment, lateral; *e*, rostrum, anterior; *f*, antenna; *g*, mandible; *h*, maxillule; *i*, distal part of maxilla; *j*, maxilliped; *k*, first leg, anterior; *l*, second leg, posterior; *m*, third leg, posterior; *n*, basipod of fourth leg, posterior; and *o*, fifth pair of legs, posterior.

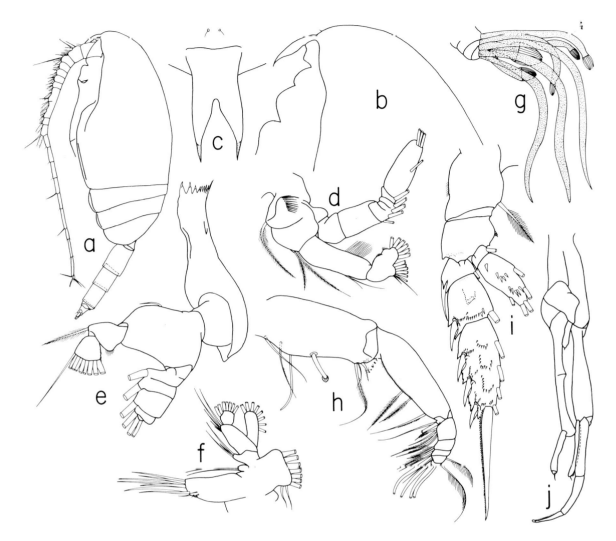

Fig. 9. *Scolecithricella schizosoma*, n. sp., male: *a*, habitus, lateral; *b*, forehead, lateral; *c*, rostrum, anterior; *d*, antenna; *e*, mandible; *f*, maxillule; *g*, distal part of maxilla; *h*, maxilliped; *i*, second leg, posterior; and *j*, fifth pair of legs, anterior.

spiniform process along external margin as in *S. dentata*. Second leg coxa with a conspicuous patch of spinules on external edge. Second exopodal segment in second and third legs with a large central patch of spinules on posterior surface. Fifth leg (Figure 8*o*) 1-segmented, elongated lamelliform, smoothly curved inward, attached to common basal coupler, with a small distal spine and large inner spine. A minute spine usually found next to distal spine and also on external edge.

Male. Prosome length, 1.56–1.64 mm; body length, 2.10–2.28 mm. Similar in habitus to female, but body (Figure 9*a*) slender, and distal margin of metasome smoothly rounded without an incision. Urosome about $^2/_5$ length of prosome. Second urosomal segment longest. Third and fourth urosomal segments of about equal length, about $^3/_4$ length of second urosomal segment. Rostral rami (Figure 9*c*) markedly divergent, and distal filaments slightly curved inward.

Antennule (Figure 9*a*) reaching distal end of second urosomal segment, with eighth through twelfth segments and twentieth and twenty-first segments fused, respectively, as in *S. minor* and *S. vittata*. Antenna (Figure 9*d*) and mandible (Figure 9*e*) agree in meristic details with those of *S. minor*.

Maxillule (Figure 9*f*) with number of setae on various lobes identical with those of female, but setae on inner lobes and basis much reduced in size. Maxilla (Figure 9*g*) similar to that of female, setae being reduced only slightly in development, but 1 of 2 short brush-form sensory filaments considerably larger than in female. Maxilliped endopod (Figure 9*h*) with 4, 2, 1, 2 + 1, and 3 + 1 setae on first through fifth segments.

First to fourth legs (Figure 9*i*) agree in all details with those of female, including endopods, which have no spiniform process on distal end. Fifth pair of legs (Figure 9*j*) similar to that of *S. dentata* male as described by Giesbrecht [1892] (as *S. dubia*) and Rose [1942] except for third exopodal segment of left leg, which is curved in *dentata* but straight in *schizosoma.* The fifth pair of legs in these two species is readily distinguishable from that of other *Scolecithricella* males by the following characters: right endopod moderately developed, right exopod and left endopod each bearing a minute seta distally, and both reaching distal end of first exopodal segment of left leg.

Remarks. *Scolecithricella schizosoma* is most closely related to *S. dentata* but can be readily distinguished from it in the female by the large body size, the absence of a distal spiniform process on the leg endopods, and the elongate and curved fifth legs. According to Rose [1942] the male of *S. dentata* differs from that of *S. schizosoma* in having a distal spiniform process on the leg endopod.

Scolecithricella schizosoma was represented in the study by 140 females and 11 males from 26 widely distributed stations ranging from the edge of the Antarctic to as far north as 38°25′S.

The species name *schizosoma* refers to the female body, which has a characteristic incision on the posterior edge of the metasome.

Type specimens selected from the specimens obtained from *Eltanin* cruise 46, station 9, 1000–0 m, have been deposited in the U.S. National Museum (USNM) of Natural History. Female holotype, USNM catalog no. 170760; male allotype, USNM catalog no. 170761.

Scolecithricella dentipes Vervoort, 1951
Figs. 10 and 11

Scolecithricella dentipes Vervoort, 1951, p. 103, figs. 55–59; 1957, p. 103, figs. 88–92.

Scolecithricella robusta; Vervoort, 1957, p. 105, figs. 92–95.

Occurrence. The following station list shows the occurrence of *S. dentipes* Vervoort, 1951:

Eltanin Cruise 17

Sta. 18, 768–0 m, 3F (2.52–2.68 mm)
Sta. 20, 768–0 m, 1F (2.64 mm)
Sta. 26, 2560–0 m, 8F (2.56–2.76 mm); 1M (2.88 mm)
Sta. 41, 625–0 m, 2F (2.60–2.68 mm)
Sta. 52, 1052–0 m, 10F
Sta. 54, 684–0 m, 7F (2.64–2.72 mm);
            1M (2.88 mm)
Sta. 56, 1251–0 m, 6F (2.64–2.72 mm);
            1M (2.92 mm)
Sta. 59, 1251–0 m, 9F (2.44–2.56 mm);
            1M (2.68 mm)
Sta. 62, 1251–0 m, 5F (2.68–2.76 mm)
Sta. 63, 1251–0 m, 17F (2.56–2.76 mm);
            6M (2.88–3.00 mm)
Sta. 69, 3146–0 m, 7F (2.56–2.64 mm);
            4M (2.88–3.00 mm)
Sta. 77, 836–0 m, 44F (2.56–2.76 mm);
            9M (2.88–2.92 mm)
Sta. 80, 625–0 m, 47F (2.44–2.60 mm);
            6M (2.72–2.84 mm)
Sta. 85, 625–0 m, 1F (2.60 mm)
Sta. 88, 2502–0 m, 3F (2.60 mm)

Eltanin Cruise 21

Sta. 252, 1050–0 m, 1F (2.60 mm)
Sta. 257, 1000–0 m, 20F (2.36–2.64 mm);
            12M (2.80–2.88 mm)
Sta. 264, 1230–0 m, 53F (2.40–2.68 mm);
            7M (2.84–2.88 mm)
Sta. 272, 1000–0 m, 3F (2.52 mm); 2M

Eltanin Cruise 22

Sta. 1503, 2505–0 m, 1F
Sta. 1528, 2452–0 m, 1F (2.72 mm)

Eltanin Cruise 23

Sta. 1697, 2274–0 m, 2F (2.56 mm);
            1M (2.80 mm)
Sta. 1700, 1275–0 m, 6F (2.56 mm)
Sta. 1704, 800–0 m, 7F (2.64–2.84 mm)
Sta. 1710, 900–0 m, 11F (2.72–2.84 mm)

Eltanin Cruise 26

Sta. 1825, 1625–0 m, 2F (2.72 mm)
Sta. 1835, 1375–0 m, 1F (2.64 mm)
Sta. 1839, 3750–0 m, 5F (2.52–2.64 mm)

Eltanin Cruise 32

Sta. 1992, 3660–0 m, 1F (2.60 mm)
.Sta. 2111, 1830–0 m, 1F (2.64 mm)

Eltanin Cruise 35

Sta. 2260, 1200–0 m, 1F (2.52 mm)
Sta. 2264, 1200–0 m, 7F (2.60–2.80 mm);
　　　　2M (2.88 mm)
Sta. 2279, 1200–0 m, 1F (2.76 mm)
Sta. 2285, 1250–0 m, 19F (2.40–2.60 mm);
　　　　4M (2.72–2.92 mm)
Sta. 2289, 1200–0 m, 6F (2.48–2.68 mm)
Sta. 2293, 1300–0 m, 4F (2.48–2.68 mm);
　　　　1M (2.96 mm)
Sta. 2301, 900–0 m, 7F (2.52–2.68 mm);
　　　　1M (2.96 mm)

Eltanin Cruise 46

Sta. 2,　1000 m, 11F (2.60–2.72 mm)
Sta. 4,　500 m, 43F (2.48–2.68 mm),
　　　　56M (2.84–29.2 mm)
　　　1000 m, 123F (2.46–2.72 mm);
　　　　30M (2.80–2.92 mm)
Sta. 5,　500 m, 3M (2.96 mm)
　　　1000 m, 323F (2.56–2.68 mm);
　　　　32M (2.84–2.92 mm)
Sta. 6,　1000 m, 128F (2.56–2.68 mm);
　　　　7M (2.88–3.00 mm)
Sta. 7,　500 m, 10F (2.44–2.68 mm);
　　　　2M (2.92 mm)
Sta. 8,　500–0 m, 29F (2.48–2.68 mm);
　　　　1M (2.92 mm)
Sta. 9,　500 m, 1F (2.60 mm)
　　　1000–0 m, 79F (2.40–2.72 mm);
　　　　5M (2.80–2.92 mm)
　　　1000 m, 2F (2.52–2.76 mm);
　　　　1M (2.96 mm)
Sta. 10, 1000–0 m, 101F (2.48–2.76 mm);
　　　　12M (2.84–2.88 mm)
Sta. 11, 1000 m, 31F (2.48–2.72 mm);
　　　　2M (2.88 mm)
Sta. 15, 1000 m, 18F (2.48–2.64 mm)
Sta. 16,　500 m, 1F (2.64 mm);

　　　　5M (2.84–2.92 mm)
　　1000 m, 59F (2.48–2.68 mm)
Sta. 17, 500 m, 14F (2.52–2.72 mm);
　　　　6M (2.84–2.92 mm)
　　731 m, 52F (2.56–2.76 mm);
　　　　26M (2.80–2.92 mm)

Atlantis II Cruise 31

Sta. RHB 1441, 190–0 m, 1F (2.60 mm)

Total: 1356F and 247M

Female. Prosome length, 2.00–2.32 mm; body length, 2.36–2.84 mm. Body strongly built. Cephalosome and first metasomal segment fused. Fourth and fifth metasomal segments fused. In lateral view (Figures 10*a* and 10*b*), last metasomal segment produced distally, covering anterior $1/3$ of genital segment. Dorsal margin of last metasomal segment with a large incurvation at level of joint between metasome and urosome.

Urosome about $1/4$ length of prosome. Genital segment longer than second or third urosomal segment by $1/3$ its length, in lateral view (Figure 10*d*), with large round genital swelling. Lateral skeletal plate of genital orifice large, elongate. Spermatheca with long, thin fingerlike vesicle extending dorsad. Rostrum (Figure 10*e*) with long, strong rami each bearing a short, soft, and transparent sensory filament.

Antennule (Figure 10*a*) with 24 free segments, eighth and ninth segments being fused, reaching as far as distal end of caudal ramus. In antenna (Figure 10*f*), second exopodal segment with a small distal seta; seventh exopodal segment with a small middle seta in addition to 3 distal setae. Mandible (Figure 10*g*) with 3 setae on basis, the middle one of which is very small, and 2 setae on first endopodal segment. Maxillule (Figure 10*h*) with 2 posterior and 10 distal setae on first, 2 setae on second, and 4 setae on third inner lobe; 5 setae on basis, 3 + 5 setae on endopod, 9 setae on exopod, and 9 setae on outer lobe. Maxilla (Figure 10*i*) with 3 setae on each of first 4 lobes, 3 setae and 1 vermiform sensory filament on fifth lobe, and 5 brush-form and 3 vermiform sensory filaments on endopod. Maxilliped endopod (Figure 10*j*) with 4, 3, 3, 3 + 1, and 4 setae on first through fifth segments.

In first leg (Figure 10*k*), each exopodal segment with an outer spine. In second leg (Figure 10*l*), outer spine on first exopodal segment curved and much

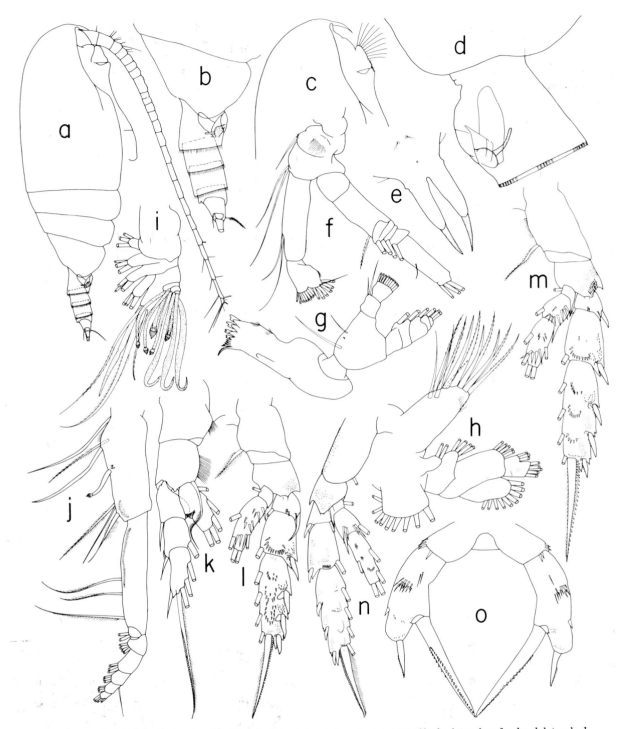

Fig. 10. *Scolecithricella dentipes* female: *a*, habitus, lateral; *b*, posterior part of body, lateral; *c*, forehead, lateral; *d*, genital segment, lateral; *e*, rostrum, anterior; *f*, antenna; *g*, mandible; *h*, maxillule; *i*, maxilla; *j*, maxilliped; *k*, first leg, anterior; *l*, second leg, posterior; *m*, third leg, posterior; *n*, fourth leg, posterior; and *o*, fifth pair of legs, posterior.

longer than outer spines of second and third exopodal segments. In third leg (Figure 10*m*), outer spine of first exopodal segment smaller than outer spines of second and third exopodal segments. Fourth leg exopod (Figure 10*n*) with all outer spines of equally small size. Endopods of first to fourth legs without distal spiniform process along external margin. Posterior surface of second to fourth legs armed with

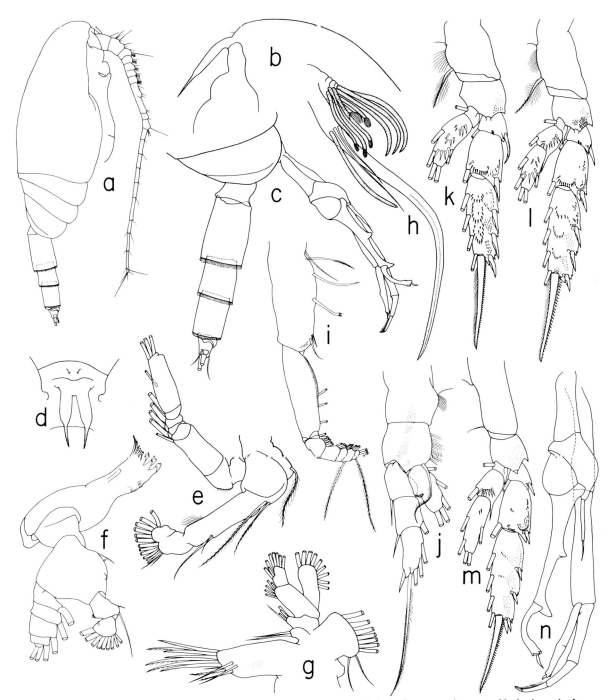

Fig. 11. *Scolecithricella dentipes* male: *a*, habitus, lateral; *b*, forehead, lateral; *c*, posterior part of body, lateral; *d*, rostrum, anterior; *e*, antenna; *f*, mandible; *g*, maxillule; *h*, distal part of maxilla; *i*, maxilliped; *j*, first leg, anterior; *k*, second leg, posterior; *l*, third leg, posterior; *m*, fourth leg, posterior; and *n*, fifth pair of legs, anterior.

spines and spinules. Spines on third and fourth legs arranged in arcs.

Fifth leg (Figure 10*o*) 2-segmented. Distal segment elongate, curved inward, with a tiny external spine, a small distal spine, and a large internal spine. Internal spine with external edge conspicuously serrated, longer than segment itself, and nearly 3 times as long as distal spine. Posteriorly, first segment armed with large spines along laterodistal margin; second segment with a patch of large spines midway and some spinules distal to internal spine.

Male. Prosome length, 1.96–2.16 mm; body

length, 2.72–3.00 mm. Similar in meristic details of appendages to female but in habitus to male of *S. schizosoma* described above. Fourth and fifth metasomal segments partially separated on dorsal side by a clearly visible articulation suture (Figures 11*a* and 11*c*). Urosome about ²/₅ length of prosome. Second urosomal segment longest, about 1.5 times length of third. Fourth segment only slightly longer than third. Rostrum (Figure 11*d*) similar to that of female.

Antennule (Figure 11*a*) reaches about distal end of second urosomal segment, with eighth through twelfth segments fused. Twentieth and twenty-first segments fused in right antennule but separate in left. In antenna (Figure 11*e*), distal seta of second and middle seta of seventh exopodal segment well developed. Mandible (Figure 11*f*) with masticatory blade somewhat reduced. Basis with 2 small setae and a small conical or wartlike projection. First endopodal segment with 2 setae. Maxillule (Figure 11*g*) with same meristic details as that of female but somewhat reduced, particularly in development of inner lobes. Maxilla (Figure 11*h*) and maxilliped (Figure 11*i*) also same in meristic details as those of female. In maxilliped, 3 distal setae of coxa greatly reduced in size, but some endopodal setae much better developed than in female.

First to fourth legs (Figures 11*j*–11*m*) similar in all anatomical details to those of female. In right fifth leg (Figure 11*n*), endopod moderately developed; first and second exopodal segments almost completely fused. Second exopodal segment with a large distal process medially. Third exopodal segment strongly curved inward, with a small degenerate seta distally. Left fifth leg with 1-segmented endopod and 3-segmented exopod. Endopod barely reaches distal end of second exopodal segment.

Remarks. *Scolecithricella dentipes* was originally described by Vervoort [1951] from female specimens 2.60–2.70 mm long collected from 66°S, 11°W in the Atlantic sector of the Antarctic. Vervoort [1957] found the species at a number of localities between 47°43′S and 65°27′S and between 45°32′E and 116°42′E in the Indian Ocean sector of the Antarctic.

Vervoort [1957] recorded *Scolecithricella robusta* (T. Scott, 1894) from between 45°10′S and 66°05′S and between 73°50′E and 152°46′E in the Indo-Pacific sector of the Antarctic and distinguished the species from *S. dentipes* by minor differences in the fifth pair of legs, namely, the shape of the distal segment, the size and number of teeth along the external margin of the internal spine, and the number of spinules on the posterior surface of the leg. In the present study a total of 1356 females has been examined, and their fifth pairs of legs were found to be quite variable. The fifth pairs of legs as figured by Vervoort [1951, 1957] for *S. dentipes* and *S. robusta* were well within the range of variation shown by the present specimens. It is therefore believed that the females that Vervoort [1951, 1957] referred to *S. dentipes* and *S. robusta* belong to the same species.

Scolecithricella robusta was originally described as *Amallophora robusta* by T. Scott [1894] from female specimens 3 mm long obtained in the Gulf of Guinea. It has been redescribed briefly as *Scolecithrix robusta* by Farran [1908] from specimens 2.65–3.10 mm long taken off the west coast of Ireland and by With [1915] as *Scaphocalanus robustus* from a specimen 2.78 mm long collected from the northern North Atlantic. According to the descriptions by these authors the fifth leg of *S. robusta* is not armed with spinules, and its inner spine is relatively short, in contrast to that of *S. dentipes* or *S. robusta* as described by Vervoort [1957]. all specimens referred to *S. dentipes* in the present study are very characteristic in having a long, thin fingerlike vesicle of the spermatheca. The spermatheca of *S. robusta* has not been described.

The male of *S. dentipes* has been described by Vervoort [1957] from specimens 2.79 mm long collected from the Indian Ocean sector of the Antarctic. The present specimens of both the female and the male agreed well with Vervoort's [1951, 1957] original descriptions. However, the species is redescribed fully here to provide more reliable diagnostic characters.

Scolecithricella dentipes was the most common deep-living *Scolecithricella* species throughout the study area. Altogether, 1356 females and 247 males were found, which represent about 30% of the total *Scolecithricella* found in the present study.

Scolecithricella parafalcifer, n. sp.
Fig. 12

Occurrence. The following station list shows the occurrence of *S. parafalcifer*, n. sp.:

Eltanin Cruise 17

Sta. 26, 2560–0 m, 1F
Sta. 88, 2502–0 m, 3F (2.14 mm)

***Eltanin* Cruise 23**

Sta. 1700, 1275–0 m, 5F (1.86–2.04 mm)

***Eltanin* Cruise 26**

Sta. 1835, 1375–0 m, 1F (2.02 mm)
Sta. 1842, 1350–0 m, 2F (1.98 mm)

***Eltanin* Cruise 35**

Sta. 2279, 1200–0 m, 4F (1.94–2.00 mm)
Sta. 2285, 1250–0 m, 3F (1.98–2.00 mm)

Total: 19F

Female. Prosome length, 1.54–1.84 mm; body length, 1.86–2.14 mm. Similar in habitus to *S. dentipes*, but posterior end of metasome less produced distad, covering only anterior 1/4 of genital segment. Lateral skeletal plate of genital orifice short and wide (Figures 12*b* and 12*c*), with broadly rounded distal end. Spermatheca with long, curved distal vesicle, which is similar to but much thicker than that in *S. dentipes*. Rostrum (Figure 12*d*) and all appendages (Figures 12*e*–12*i*) of cephalosome agree in details with those of *S. dentipes*. First to fourth legs (Figures 12*j*–12*l*) also similar to those of *S. dentipes* except for pattern of spinous armature on posterior surface as described below. In second leg (Figure 12*k*), all segments bearing spines or spinules or both. Second endopodal segment with spines clearly arranged in arcs. Second exopodal segment with a large patch of spines centrally. Third exopodal segment with a round patch of spines in anterior 1/3 of segment. Outer spines of exopod relatively large. Third leg (Figure 12*l*) also bearing spinous armature on all segments. Spines on exopodal segments in round patches instead of arcs.

Fifth leg (Figure 12*m*) similar to that of *S. dentipes*, but without spines on posterior surface. Internal spine shorter than segment and about 3 times as long as distal spine. External spine small, located at same level as internal spine.

Remarks. This species is close to *Scolecithrix falcifer* Farran, 1926, described from a single female specimen (2.0 mm long) captured in the Bay of Biscay and redescribed as *Amallothrix falcifer* by Roe [1975] from specimens (1.82–2.20 mm long) taken in the northeastern Atlantic (at 28°N, 14°W and 18°N, 25°W). According to these descriptions and my examination (T. Park, unpublished data,

1977) of specimens from the Gulf of Mexico that agreed with Roe's [1975] description, Farran's *falcifer* differs from *parafalcifer* in the fifth pair of legs and mandible. In the *falcifer* fifth leg the external spine is located midway between the internal and distal spines, the distal spine is only slightly shorter than the internal spine, and the posterior surface is armed with spinules. The basis of the *falcifer* mandible has three setae as in *parafalcifer*, but the posteriormost seta is the longest, which in the latter is very small.

Scolecithricella parafalcifer was represented in the present study by 19 females, all of which were taken north to the Antarctic Convergence except one, which was collected in antarctic waters (*Eltanin* cruise 17, station 26). The male was not found.

Type specimens selected from the specimens taken at station 1700 of *Eltanin* cruise 23 have been deposited in the U.S. National Museum of Natural History. Holotype, USNM catalog no. 170762.

Scolecithricella pseudopropinqua, n. sp.
Figs. 13 and 14

Occurrence. The following station list shows the occurrence of *S. pseudopropinqua,* n. sp.:

***Eltanin* Cruise 17**

Sta. 80, 625–0 m, 1F (3.20 mm)
Sta. 88, 2502–0 m, 2M (3.52–3.56 mm)

***Eltanin* Cruise 21**

Sta. 213, 1050–0 m, 2F (2.96–3.24 mm)

***Eltanin* Cruise 23**

Sta. 1700, 1275–0 m, 7F (3.20–3.32 mm)
Sta. 1704, 800–0 m, 9M (3.32–3.44 mm)
Sta. 1710, 900–0 m, 4F (3.12–3.32 mm)

***Eltanin* Cruise 26**

Sta. 1825, 1625–0 m, 3F (3.32–3.40 mm);
        1M (3.52 mm)
Sta. 1842, 1350–0 m, 1F

***Eltanin* Cruise 35**

Sta. 2279, 1200–0 m, 1F (3.20 mm)
Sta. 2285, 1250–0 m, 4F (3.12–3.32 mm)

***Eltanin* Cruise 46**

Sta. 2, 1000 m, 1M (3.32 mm)

Total: 23F and 13M

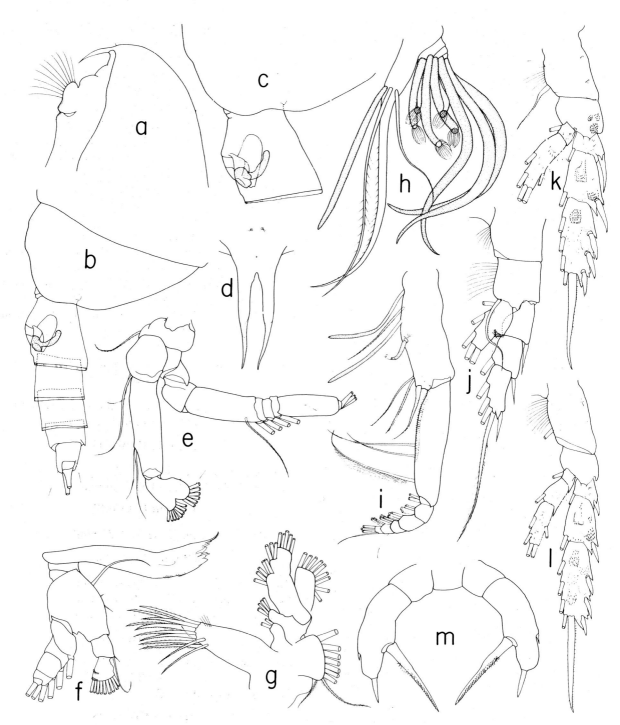

Fig. 12. *Scolecithricella parafalcifer*, n. sp., female: *a*, forehead, lateral; *b*, posterior part of body, lateral; *c*, genital segment, lateral; *d*, rostrum, anterior; *e*, antenna; *f*, mandible; *g*, maxillule; *h*, distal part of maxilla; *i*, maxilliped; *j*, first leg, anterior; *k*, second leg, posterior; *l*, third leg, posterior; and *m*, fifth pair of legs, posterior.

Female. Prosome length, 2.52–2.72 mm; body length, 2.96–3.40 mm. Similar in habitus to *S. dentipes* but readily distinguishable from it by larger body size and some features of genital segment.

Viewed laterally (Figure 13*c*), spermathecal vesicle short, thick, and bent forward to varying degrees. Genital segment with a row of spinules immediately anterior to genital field. Rostrum (Figure 13*e*) and

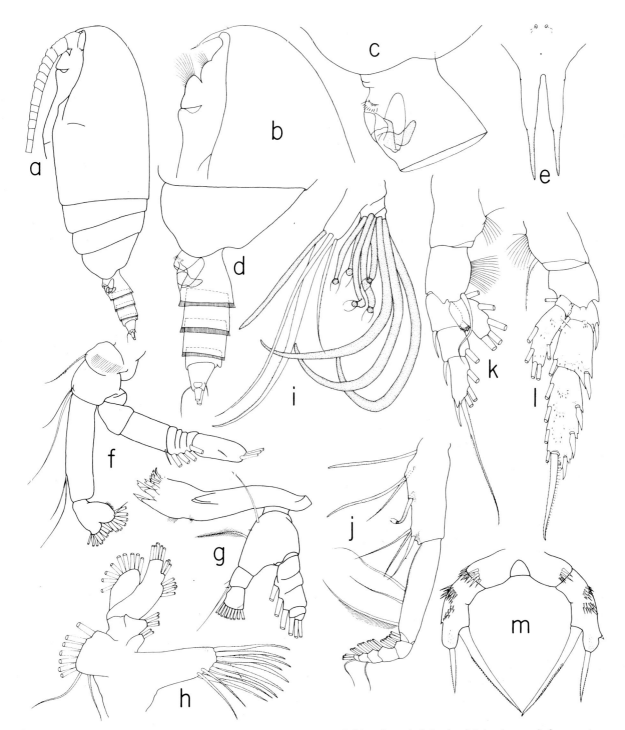

Fig. 13. *Scolecithricella pseudopropinqua*, n. sp., female: *a*, habitus, lateral; *b*, forehead, lateral; *c*, genital segment, lateral; *d*, posterior part of body, lateral; *e*, rostrum, anterior; *f*, antenna; *g*, mandible; *h*, maxillule; *i*, distal part of maxilla; *j*, maxilliped; *k*, first leg, anterior; *l*, second leg, posterior; and *m*, fifth pair of legs, posterior.

all appendages (Figures 13*f*–13*m*) similar to those of *S. dentipes* except those noted below.

Distal seta of mandibular basis (Figure 13*g*) well developed. Outer spine of first exopodal segment of second leg (Figure 13*l*) relatively short. In fifth leg (Figure 13*m*), internal spine considerably longer than segment; distal spine nearly 1/2 length of internal spine.

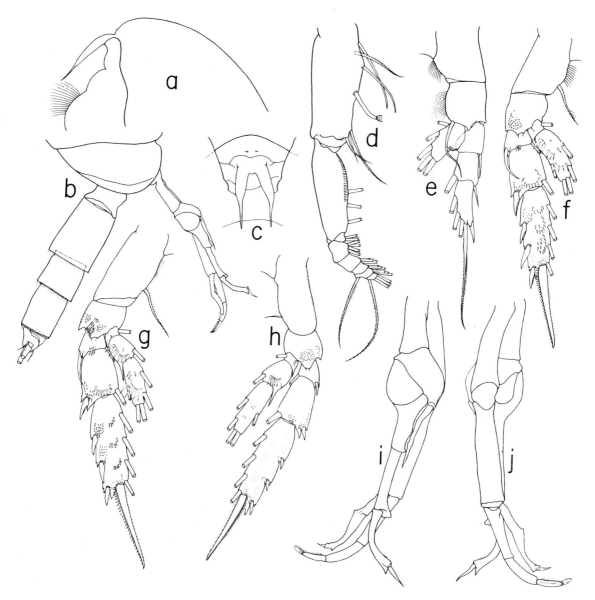

Fig. 14. *Scolecithricella pseudopropinqua*, n. sp., male: *a*, forehead, lateral; *b*, posterior part of body, lateral; *c*, rostrum, anterior; *d*, maxilliped; *e*, first leg, anterior; *f*, second leg, posterior; *g*, third leg, posterior; *h*, fourth leg, posterior; *i*, fifth pair of legs, right side; and *j*, fifth pair legs, left side.

Male. Prosome length, 2.36–2.52 mm; body length, 3.32–3.56 mm. Similar morphologically to male of *S. dentipes* but considerably larger. All head appendages, from antennule through maxilla, agree in all anatomical details with those of *S. dentipes* male. Maxilliped (Figure 14*d*) agrees in meristic details with that of *S. dentipes* male, but 3 distal setae of coxa much better developed. First to fourth legs (Figure 14*e*–14*h*) also similar to those of *S. dentipes* male except that first exopodal spine long and curved in both second and third legs instead of just in

second leg. Fifth pair of legs (Figures 14*i* and 14*j*) resembles closely that of *S. dentipes* male but differs from it in the following aspects: In right leg, distal process of second exopodal segment relatively small; third exopodal segment only slightly curved.

Remarks. This new species seems to be similar to *Scolecithricella propinqua*, which was originally described by Sars [1920] from a female specimen taken from 38°02′N, 10°44′W in the North Atlantic. Sars' specimen was indicated as being 2.30 mm long in the description but 2.90 mm long in the figures

[Sars, 1920, 1925]. Tanaka [1962] redescribed *S. propinqua*, including the male from specimens collected from off the east coast of middle Japan. His specimens were 2.95 mm long in the female and 2.98 mm long in the male. Vervoort [1965] recorded the species from the Gulf of Guinea on the basis of specimens 2.80–3.60 mm long in the female and 3.04–3.41 mm long in the male, which were said to be identical with Tanaka's [1962].

There is some disagreement between the original descriptions of the species by Sars [1920, 1925] and its redescription by Tanaka [1962]; namely, the fifth leg is bimerous in Sars' specimen but trimerous in Tanaka's, and a row of spinules in front of the genital field shown by Tanaka is not indicated in Sars' description. However, Sars' description is too general to permit detailed comparisons. In the present female specimens the fifth pair of legs was similar to that figured by Sars [1925], but the genital segment had a row of spinules in front of the genital field, as shown by Tanaka [1962].

The males described here are believed to belong to this species because of their similarity in body size, meristic details of the appendages, and distribution. They seem to be closely related to the *S. propinqua* male as described by Tanaka [1962] but seem to differ from it in the size of the first exopodal spines of the second and third legs and in the size of the third exopodal segment and its distal seta of the right fifth leg.

In the present study the new species was represented by 23 females and 13 males which were found at 11 stations widely distributed in waters north to the Antarctic Convergence. Type specimens selected from the specimens from stations 1700 and 1704 of *Eltanin* cruise 23 have been deposited in the U.S. National Museum of Natural History. Female holotype, USNM catalog no. 170763; male allotype, USNM catalog no. 170764.

Scolecithricella valida (Farran, 1908)
Figs. 15 and 16

Scolecithrix valida Farran, 1908, p. 55 pl. 5, figs. 14–17, pl. 6, fig. 7; 1929, p. 244.
Amallothrix valida; Sewell, 1929, p. 217, fig. 80.—Brodsky, 1950, p. 260, figs. 169, 170.
Scaphocalanus validus; With, 1915, p. 198, pl. 7, fig. 11, text fig. 62.—Wilson, 1932, p. 78, fig. 53.
Scolecithricella valida; A. Scott, 1909, p. 92, pl. 32, figs. 1–9.—Vervoort, 1957, p. 107.—Tanaka, 1962, p. 70, fig. 143.

Occurrence. The following station list shows the occurrence of *S. valida* (Farran, 1908):

Eltanin Cruise 17

Sta. 26, 2560–0 m, 2F (4.04–4.08 mm)
Sta. 54, 684–0 m, 2F (4.04–4.08 mm)
Sta. 59, 1251–0 m, 1F (3.84 mm)
Sta. 62, 1251–0 m, 1F (4.00 mm)
Sta. 69, 3146–0 m, 2F (3.84–3.88 mm)
Sta. 88, 2502–0 m, 1F (3.84 mm)

Eltanin Cruise 22

Sta. 1528, 2452–0 m, 1F (4.28 mm)

Eltanin Cruise 23

Sta. 1685, 2250–0 m, 1F (4.00 mm)
Sta. 1697, 2274–0 m, 4F (3.80–4.00 mm)
Sta. 1710, 900–0 m, 1M (4.04 mm)

Eltanin Cruise 26

Sta. 1825, 1625–0 m, 3F (4.04 mm)
Sta. 1839, 3750–0 m, 1F
Sta. 1842, 1350–0 m, 1F (3.92 mm)

Eltanin Cruise 32

Sta. 1992, 3660–0 m, 2F (3.84–4.08 mm)
Sta. 2111, 1830–0 m, 1F (3.80 mm)
Sta. 2133, 1829–0 m, 1F (4.40 mm)

Eltanin Cruise 35

Sta. 2279, 1200–0 m, 1M (4.04 mm)
Sta. 2285, 1250–0 m, 2F (3.92 mm);
                2M (4.00–4.04 mm)

Eltanin Cruise 46

Sta. 6, 1000 m, 5F (3.84–3.96 mm)
Sta. 9, 1000–0 m, 1F (4.12 mm)
         1000 m, 1F (4.04 mm)
Sta. 10, 1000–0 m, 19F (3.80–4.36 mm)
Sta. 11, 1000 m, 4F (3.96–4.12 mm)
Sta. 15, 1000 m, 11F (3.80–4.04 mm)
Sta. 16, 1000 m, 3F (3.84–3.88 mm)

Total: 70F and 4M

Female. Prosome length, 3.08–3.40 mm; body length, 3.80–4.40 mm. Similar morphologically to *Scolecithricella dentipes, S. pseudopropinqua,* and *S. parafalcifer* described above but readily dis-

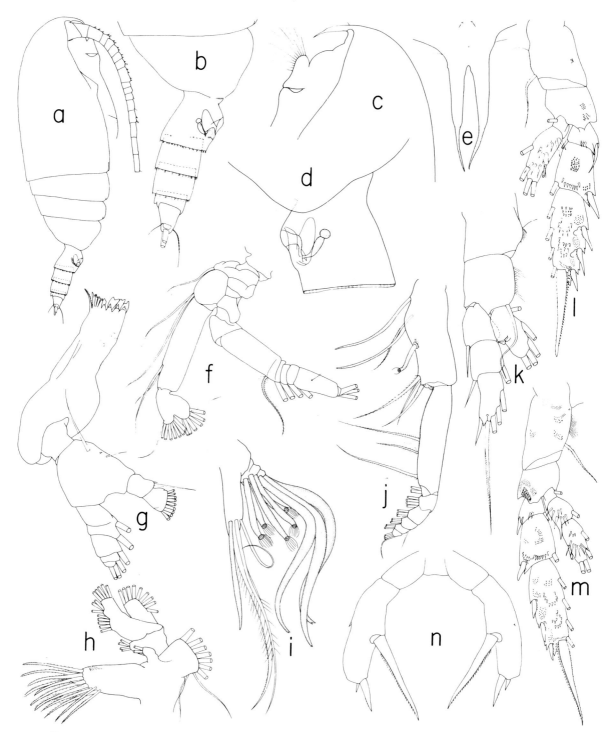

Fig. 15. *Scolecithricella valida* female: *a*, habitus, lateral; *b*, posterior part of body, lateral; *c*, forehead, lateral; *d*, genital segment, lateral; *e*, rostrum, anterior; *f*, antenna; *g*, mandible; *h*, maxillule; *i*, distal part of maxilla; *j*, maxilliped; *k*, first leg, anterior; *l*, second leg, posterior; *m*, third leg, posterior; and *n*, fifth pair of legs, posterior.

tinguishable from them by the following diagnostic characters: Body considerably larger. Spermathecal vesicle (Figure 15*d*) elongate, with a round distal bursa distinctly set off from rest of vesicle. Lateral

skeletal plate of genital orifice somewhat triangular and relatively small.

Rostrum (Figure 15*e*) and cephalosomal appendages from antennule through maxilliped (Figures

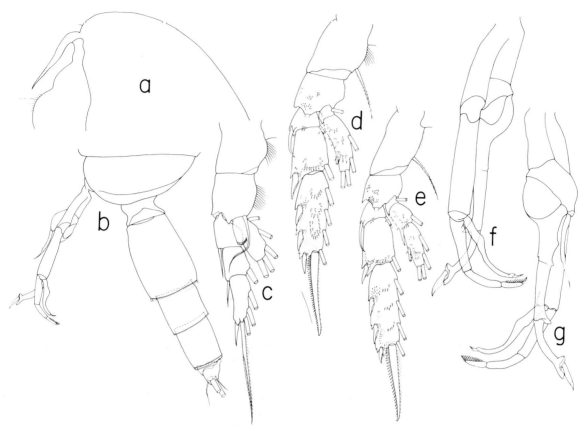

Fig. 16. *Scolecithricella valida* male: *a*, forehead, lateral; *b*, posterior part of body, lateral; *c*, first leg, anterior; *d*, second leg, posterior; *e*, third leg, posterior; *f*, fifth pair of legs, posterior; and *g*, fifth pair of legs, right side.

15*f*–15*j*) agree in all meristic details with those of *S. dentipes*. However, seta on second exopodal segment of antenna (Figure 15*f*) relatively well developed. In mandible (Figure 15*g*), 1 of 2 setae on first endopodal segment very small. First to third legs (Figures 15*k*–15*m*) also similar to those of *S. dentipes* except patterns of spinous armature. In second leg (Figure 15*l*), basis with 3 distinct patches of spinules; second exopodal segment with a central patch of spinules. In third leg (Figure 15*m*), coxa with several patches of spinules; second exopodal segment with a patch of spinules centrally. Fourth leg was not available for observation.

Fifth leg (Figure 15*n*) 2-segmented, elongate, curved inward, with a tiny external spine, a small and a medium-sized distal spine, and a large internal spine. External and internal spines located approximately in middle of segment. Internal spine about 2.5 times as long as large distal spine and shorter than segment itself. No spines or spinules found on posterior surface.

Male. Prosome length, 2.80–2.84 mm; body length, 4.00–4.04 mm. Similar morphologically to males of *S. dentipes* and *S. pseudopropinqua* described above but distinguishable from them by its large body and details of fifth pair of legs.

Rostrum and head appendages from antennule through maxilla agree in all meristic details with those of *S. dentipes*. Maxilliped with well-developed distal setae on coxa as in *S. pseudopropinqua*. First to third legs (Figures 16*c*–16*e*) very similar to those of *S. pseudopropinqua*. First exopodal spines in both second and third legs long and curved.

Fifth pair of legs relatively short, reaching distal end of third urosomal segment (Figure 16*b*). Right leg (Figures 16*f* and 16*g*) with relatively well developed endopod. Third exopodal segment of right leg only slightly curved, with a long distal process along external margin and a relatively well developed distal seta. In left leg, endopod long; in natural condition, reaching close to distal end of exopod.

Remarks. *Scolecithricella valida* was originally described as *Scolecithrix valida* by Farran [1908] from female specimens 3.8–3.9 mm long collected from deep waters off the west coast of Ireland. Sub-

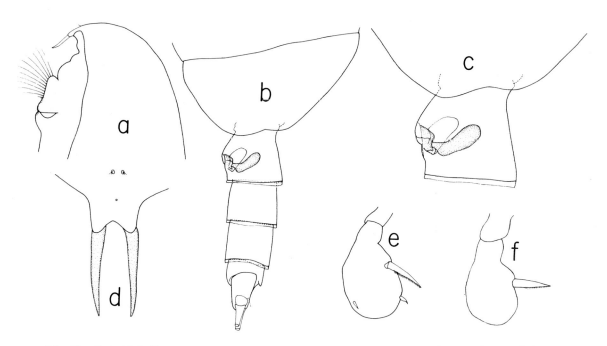

Fig. 17.   *Scolecithricella ovata* female: *a*, forehead, lateral; *b*, posterior part of body, lateral; *c*, genital segment, lateral; *d*, rostrum, anterior; and *e*, *f*, fifth legs, posterior.

sequently, Farran [1929] recorded the species from 71°41′S, 166°47′W in the Pacific sector of the Antarctic on the basis of two female specimens 4.06 mm in body length.

In the Atlantic Ocean the species has been recorded from 61°30′N, 17°08′W [With, 1915], south of Martha's Vineyard [Wilson, 1932], and 27°36′N, 38°92′W [Sars, 1925]. However, Sars' specimen was only 2.1 mm long and showed some morphological differences from Farran's original descriptions. As Farran [1929] has already pointed out, Sars' specimen does not seem referable to *S. valida*. In the Pacific Ocean the species has been reported from 31°54′S, 88°17′W [Wilson, 1942], the Sea of Okhotsk and the Bering Sea [Brodsky, 1950], and off the east coast of middle Japan [Tanaka, 1962]. Wilson's [1942] record is accompanied by two figures of the female fifth leg, which are not only different from each other but also different from the leg as figured originally by Farran [1908].

The species has also been recorded from 10°26′N, 74°32.5′E in the Arabian Sea [Sewell, 1929] 00°17.6′S, 129°14.5′E and 03°58′S, 128°20′E in the Malay Archipelago [A. Scott, 1909], 66°05′S, 73°50′E in the Indian Ocean sector of the Antarctic, and 44°05′S, 147°35′E south of Australia [Vervoort, 1957].

Although they are brief, Farran's [1908] original

species description and With's [1915] description based on specimens from near the type locality seem to agree well with the specimens examined in the present study. The male described here is believed to belong to this species because of its similarity in body size, meristic details of the appendages, and occurrence. It seems identical with the male described by Tanaka [1962] as a new species under the name *Scolecithricella lanceolata*. However, it is different from the males referred to *S. valida* by Wilson [1932], Brodsky [1950], and Tanaka [1962], which also differ significantly from each other.

In the present study the species was represented by 70 females and 4 males, which were found in samples from 24 stations widely distributed throughout the study area.

Scolecithricella ovata (Farran, 1905)
Fig. 17

Scolecithrix ovata Farran, 1905, p. 37, pl. 6, figs. 13–18, pl. 7, figs. 1–5.
Scolecithricella ovata; Brodsky, 1950, p. 269, fig. 179.—Vervoort, 1957, p. 102,—Tanaka, 1962, p. 55, fig. 137.—Park, 1968, p. 555, pl. 8, figs. 17–21.—Von Vaupel-Klein, 1970, p. 20.—Minoda, 1971, p. 31, pl. 2, figs. 1–10.
For earlier synonyms, see Vervoort [1957].

Occurrence. The following station list shows the occurrence of *S. ovata* (Farran, 1905):

Eltanin Cruise 17

Sta. 71, 457-0 m, 1F (2.18 mm)
Sta. 77, 836-0 m, 1F (2.18 mm)
Sta. 79, 386-0 m, 1F (2.08 mm)
Sta. 80, 625-0 m, 5F (2.12-2.18 mm)
Sta. 82, 313-0 m, 1F (2.08 mm)
Sta. 85, 625-0 m, 2F (2.12 mm)

Eltanin Cruise 21

Sta. 240, 2470-0 m, 1F
Sta. 252, 1050-0 m, 1F
Sta. 257, 1000-0 m, 5F (2.08-2.14 mm)
Sta. 264, 1230-0 m, 5F (2.10-2.18 mm)

Eltanin Cruise 23

Sta. 1697, 2274-0 m, 2F (2.04-2.24 mm)
Sta. 1700, 1275-0 m, 3F (1.78 mm)
Sta. 1704, 800-0 m, 1F (1.98 mm)

Eltanin Cruise 26

Sta. 1835, 1375-0 m, 3F (1.82 mm)
Sta. 1839, 3750-0 m, 1F

Eltanin Cruise 35

Sta. 2285, 1250-0 m, 3F (1.76-1.78 mm)

Eltanin Cruise 46

Sta. 2, 500 m, 9F (1.76-1.94 mm)
1000 m, 2F (1.86-1.90 mm)
Sta. 4, 500 m, 9F (2.08-2.20 mm)
1000 m, 5F (2.02-2.08 mm)
Sta. 5, 500 m, 7F (2.08-2.20 mm)
1000 m, 4F (2.12 mm)
Sta. 16, 500 m, 1F (2.08 mm)
Sta. 17, 500 m, 2F (2.16-2.28 mm)
731 m, 1F (2.34 mm)

Total: 76F

Female. Prosome length, 1.50-1.92 mm; body length, 1.76-2.34 mm. Similar in habitus and all meristic details of appendages to *Scolecithricella cenotelis*, a new species described below, but distinguishable from it by shape of spermatheca. Viewed laterally, spermathecal vesicle large, elongate, lying obliquely about in parallel with lateral skeletal plate of genital orifice and more or less dilated distally (Figure 17c).

Remarks. *Scolecithricella ovata* was originally described as *Scolecithrix ovata* by Farran [1905] from a female specimen 2.3 mm long captured from 53°58'N, 12°28'W off the west coast of Ireland. As summarized by Vervoort [1957], *S. ovata* has been known to occur commonly in deep waters of the North Atlantic, including the Mediterranean Sea. The species has also been reported from the North Pacific by Brodsky [1950], Tanaka [1962], Park [1968], Von Vaupel-Klein [1970], and Minoda [1971]. In the southern hemisphere the species has been known to occur in the Great Barrier Reef [Farran, 1936], off New Zealand [Farran, 1929], and in antarctic and subantarctic waters [Farran, 1929; Vervoort, 1951, 1957].

In the present study, two species have been found that agreed in anatomical details with *S. ovata* as described by Farran [1905]. The two species, however, differed from each other in the form of the spermatheca and the distribution range. The species identified here with *S. ovata* (Farran, 1905) had a short spermathecal vesicle and occurred in small numbers mainly in waters north to the Antarctic Convergence. The other species described below as a new species (*S. cenotelis*) had a long spermathecal vesicle and occurred in relatively large numbers, mainly south to the Antarctic Convergence.

A total of 76 females of *S. ovata* were found at 21 stations distributed widely in waters on and north to the Antarctic Convergence. The male has been described by Tanaka [1962] from off the Pacific coast of middle Japan. However, no male referable to this species was found during the present study.

Scolecithricella cenotelis, n. sp.
Fig. 18

Occurrence. The following station list shows the occurrence of *S. cenotelis*, n. sp.:

Eltanin Cruise 17

Sta. 18, 768-0 m, 8F (2.06-2.34 mm)
Sta. 20, 768-0 m, 3F (2.10-2.12 mm)
Sta. 26, 2560-0 m, 4F (2.16 mm)
Sta. 41, 625-0 m, 3F (2.00-2.12 mm)
Sta. 52, 1052-0 m, 7F (2.18-2.20 mm)
Sta. 54, 684-0 m, 1F (2.12 mm)
Sta. 56, 1251-0 m, 3F (1.98-2.12 mm)
Sta. 59, 1251-0 m, 3F (2.10-2.20 mm)
Sta. 62, 1251-0 m, 3F (2.16-2.22 mm)
Sta. 63, 1251-0 m, 5F (2.10-2.18 mm)

Sta. 69, 3146-0 m, 1F
Sta. 71, 457-0 m, 1F (2.18 mm)
Sta. 77, 836-0 m, 2F (2.16 mm)
Sta. 80, 625-0 m, 1F

Eltanin Cruise 21

Sta. 257, 1000-0 m, 1F
Sta. 264, 1230-0 m, 2F (1.98-2.16 mm)
Sta. 272, 1000-0 m, 3F

Eltanin Cruise 35

Sta. 2293, 1300-0 m, 1F
Sta. 2301, 900-0 m, 1F (2.02 mm)

Eltanin Cruise 46

Sta. 4,   500 m, 32F (2.00-2.18 mm)
          1000 m, 32F (2.00-2.14 mm)
Sta. 5,   500 m, 20F (2.00-2.12 mm)
          1000 m, 45F (2.02-2.14 mm)
Sta. 6,   500 m, 46F (2.02-2.12 mm)
          1000 m, 18F (2.02-2.20 mm)
Sta. 7,   500 m, 25F (1.98-2.12 mm)
Sta. 8,   500-0 m, 52F (1.98-2.26 mm)
Sta. 9,   500 m, 33F (2.02-2.18 mm)
          1000-0 m, 177F (2.02-2.20 mm)
          1000 m, 16F (1.98-2.18 mm)
Sta. 10, 500-0 m, 122F (1.98-2.18 mm)
          1000-0 m, 90F (1.98-2.18 mm)
Sta. 11, 500 m, 111F (2.10-2.20 mm)
          1000 m, 7F (2.04-2.16 mm)
Sta. 12, 844-0 m, 1F (2.20 mm)
Sta. 15, 500 m, 11F (1.98-2.12 mm)
Sta. 16, 500 m, 4F (2.06-2.12 mm)
          1000 m, 10F (1.98-2.16 mm)
Sta. 17, 500 m, 16F (2.00-2.18 mm)
          731 m, 8F (2.00-2.14 mm)

Total: 929F

Female. Prosome length, 1.62-1.98 mm; body length, 1.96-2.34 mm. Prosome about 4.5 times as long as urosome. Posterior margin of metasome (Figure 18a) with a shallow incurvation. Genital segment nearly 1.5 times as long as second or third urosomal segment. Spermatheca (Figure 18d) with a long vesicle, curved forward to a diagonal position to segment. Vesicle with a distinct distal end appearing like an unfilled vacuole. Lateral skeletal plate of genital orifice relatively short. Rostrum (Figure 18e) with a short bifurcated basal plate; each ramus with a long delicate sensory filament.

Antennule (Figure 18a) with 24 free segments, eighth and ninth of 25 segments being fused and extending slightly beyond distal end of caudal ramus. Second exopodal segment of antenna (Figure 18f) partially fused with third and bearing a small seta. Seventh exopodal segment with a middle seta in addition to 3 distal setae. In mandible (Figure 18g), basis with 2 long plumose setae; first endopodal segment with a short and a long seta, both fringed by fine hair. Maxillule (Figure 18h) with 2 posterior and 10 distal setae on first inner lobe, 2 setae on second, and 3 setae on third; 5 setae on basis, 2 + 3 setae on endopod, 5 setae on exopod, and 9 setae on outer lobe. In some specimens, basis had only 4 setae. Maxilla (Figure 18i) with 3 setae on each of first 4 lobes, 3 setae and 1 vermiform sensory filament on fifth lobe, and 3 long vermiform and 5 short brush-form sensory filaments on endopod. Endopod of maxilliped (Figure 18j) with 4, 3, 2, 2 + 1, and 4 setae on first to fifth segments.

In first leg (Figure 18k) each exopodal segment with an outer spine; endopod without distal spiniform process along external margin. In second and third legs (Figure 18l and 18m), coxa with a conspicuous notch on external margin; internal margin expanded into a large lobe fringed by long hair and with a notch distally. Endopod without spiniform distal process along external margin. In each leg, exopod with outer spines of similar size. Spinules on posterior surface of exopod arranged in similar pattern in second and third legs. Fourth leg was multilated in all specimens. Fifth leg (Figure 18n) uniramous, 2-segmented. Proximal segment small; distal segment flat with distal part expanded into a wide, roughly circular plate, with a long inner spine and usually a small distal spine. Two fifth legs often asymmetrical owing to differences in shape of distal segment and absence of distal spine on one leg.

Remarks. *Scolecithricella cenotelis* agrees with *S. ovata* in all anatomical details except for the spermatheca. In the present study, a total of 929 females were found at 31 stations distributed throughout the antarctic seas and in waters on or immediately north to the Antarctic Convergence. The male was not found. The specific name *cenotelis* refers to the distal end of the spermathecal vesicle, which looks like an empty vacuole.

A type specimen selected from the specimens taken at station 9 on *Eltanin* cruise 46 has been deposited in the U.S. National Museum of Natural History. Holotype, USNM catalog no. 170765.

Scolecithricella emarginata (Farran, 1905)
Fig. 19

Scolecithrix emarginata Farran, 1905, p. 36, pl. 7, figs. 6–17.

Scolecithricella emarginata; Tanaka, 1962, p. 66, fig. 142.—Vervoort, 1965, p. 67.—Minoda, 1971, p. 29.

Scolecithrix polaris Wolfenden, 1911, p. 252, pl. 30, figs. 1, 2, text fig. 31.—Farran, 1929, p. 243.

Scolecithricella polaris; Vervoort, 1957, p. 106, fig. 96.

For a more complete list of *Scolecithricella emarginata* synonyms, see Vervoort [1965].

Occurrence. The following station list shows the occurrence of *S. emarginata* (Farran, 1905):

Eltanin Cruise 17

Sta. 26, 2560–0 m, 11F (4.28–4.68 mm)
Sta. 54, 684–0 m, 5F (4.20–4.40 mm)
Sta. 59, 1251–0 m, 2F (4.24–4.28 mm)
Sta. 62, 1251–0 m, 4F (4.48–4.56 mm)
Sta. 63, 1251–0 m, 2F (4.20–4.40 mm)
Sta. 80, 625–0 m, 1F (4.80 mm)
Sta. 85, 625–0 m, 1F (4.36 mm)
Sta. 88, 2502–0 m, 4F (3.92–4.08 mm)

Eltanin Cruise 21

Sta. 198, 2972–0 m, 4F (4.20–4.68 mm)
Sta. 240, 2470–0 m, 1F (4.52 mm)
Sta. 252, 1050–0 m, 1F (3.84 mm)
Sta. 257, 1000–0 m, 1F (4.32 mm)
Sta. 264, 1230–0 m, 10F (3.76–4.04 mm)
Sta. 272, 1000–0 m, 1F (4.48 mm)

Eltanin Cruise 22

Sta. 1503, 2505–0 m, 1F (4.24 mm)
Sta. 1528, 2452–0 m, 6F (4.56 mm)
Sta. 1574, 2608–0 m, 5F (4.48 mm)

Eltanin Cruise 23

Sta. 1685, 2250–0 m, 2F (3.88–4.36 mm)
Sta. 1697, 2274–0 m, 7F (4.00–4.32 mm)
Sta. 1700, 1275–0 m, 5F (3.64–4.44 mm)
Sta. 1704, 800–0 m, 3F (4.40–4.60 mm)

Eltanin Cruise 26

Sta. 1825, 1625–0 m, 2F (3.88–4.12 mm)
Sta. 1835, 1375–0 m, 6F (3.88–4.12 mm)
Sta. 1842, 1350–0 m, 1F (4.80 mm)

Eltanin Cruise 32

Sta. 1993, 1830–0 m, 2F (4.12–4.28 mm)
Sta. 2111, 1830–0 m, 3F (4.00–4.44 mm)

Eltanin Cruise 33

Sta. 2174, 1830–0 m, 1F (4.32 mm)

Eltanin Cruise 35

Sta. 2285, 1250–0 m, 16F (3.80–4.48 mm)
Sta. 2293, 1300–0 m, 3F (3.92–4.44 mm)

Eltanin Cruise 46

Sta. 4, 1000 m, 13F (3.88–4.32 mm)
Sta. 5, 1000 m, 13F (3.88–4.32 mm)
Sta. 6, 1000 m, 12F (4.04–4.40 mm)
Sta. 8, 500–0 m, 3F (4.32–4.60 mm)
Sta. 9, 1000–0 m, 33F (4.04–4.60 mm)
    1000 m, 7F (4.20–4.60 mm)
Sta. 10, 1000–0 m, 16F (4.08–4.48 mm)
Sta. 11, 1000 m, 6F (4.12–4.40 mm)
Sta. 16, 1000 m, 2F (4.32–4.36 mm)

Atlantis II Cruise 31

Sta. RHB 1440, 1295–0 m, 10F (3.80–4.04 mm

Total: 226F

Female. Prosome length, 3.12–3.92 mm; body length, 3.64–4.80 mm. Body strongly built. Cephalosome and first metasomal segment fused. Fourth and fifth metasomal segments partially separated dorsally by articulation suture. Viewed laterally (Figure 19*b*), posterior edge of metasome with a conspicuous incurvation. Urosome about ¼ length of prosome. Genital segment longer than second or third urosomal segment by ⅓ its length. Laterally, genital segment (Figure 19*c*) wider than long, with large round genital swelling. Lateral skeletal plate of genital orifice relatively short. Spermatheca with elongate vesicle, curved forward, terminating with a round sac set off by a deep constriction. Rostrum (Figure 19*e*) with strong but short rami, each bearing long, delicate sensory filament.

Antennule (Figure 19*b*) extending beyond distal end of caudal ramus by last 2 segments, with 24 free segments, eighth and ninth of 25 segments being fused. In antenna (Figure 19*f*), first exopodal segment with a conical process on internal margin; second and third exopodal segments fused; seta belonging to second exopodal segment very small;

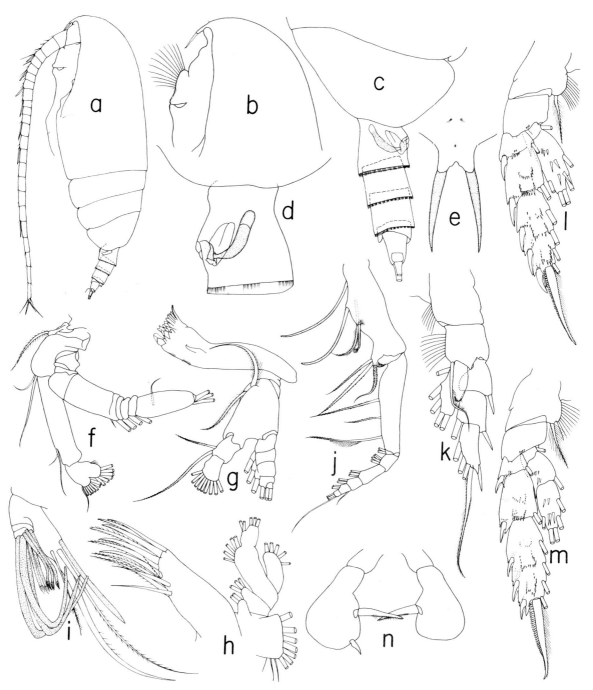

Fig. 18. *Scolecithricella cenotelis*, n. sp., female: *a*, habitus, lateral; *b*, forehead, lateral; *c*, posterior part of body, lateral; *d*, genital segment, lateral; *e*, rostrum, anterior; *f*, antenna; *g*, mandible; *h*, maxillule; *i*, distal part of maxilla; *j*, maxilliped; *k*, first leg, anterior; *l*, second leg, posterior; *m*, third leg, posterior; and *n*, fifth pair of legs, posterior.

seventh exopodal segment with a middle seta in addition to 3 distal setae. Mandible (Figure 19g) with strong masticatory blade; 2 well-developed setae on basis. Maxillule (Figure 19h) with 2 posterior and 10 distal setae on first, 2 setae on second, and 4 setae on third inner lobe; 4 setae on basis, 2 + 4 or 2 + 5

setae on endopod, 8 setae on exopod, and 9 setae on outer lobe. Maxilla (Figure 19i) with 3 setae on each of first 4 lobes, 3 setae and 1 vermiform sensory filament on fifth lobe, and 5 brush-form and 3 vermiform sensory filaments on endopod. Maxilliped (Figure 19j) with 4, 3, 3, 2 + 1, and 4 setae, in order from

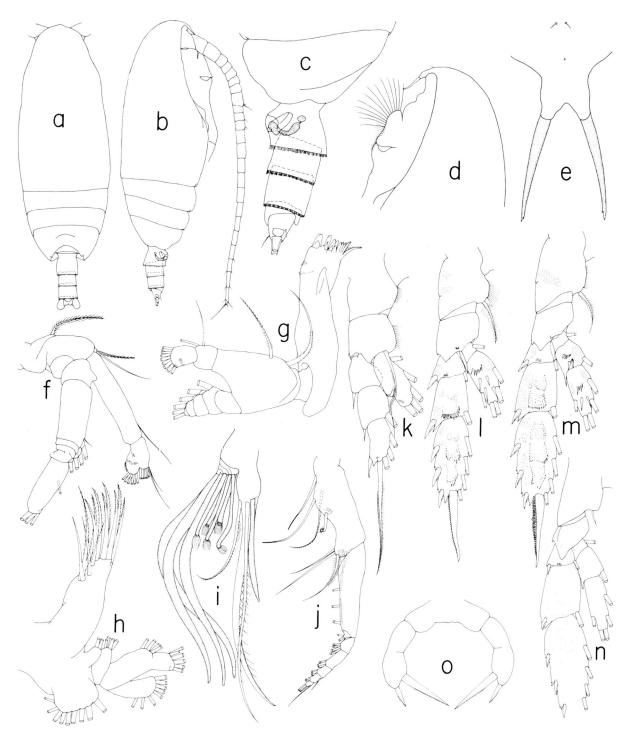

Fig. 19. *Scolecithricella emarginata* female: *a*, habitus, dorsal; *b*, habitus, lateral; *c*, posterior part of body, lateral; *d*, forehead, lateral; *e*, rostrum, anterior; *f*, antenna; *g*, mandible; *h*, maxillule; *i*, distal part of maxilla; *j*, maxilliped; *k*, first leg, anterior; *l*, second leg, posterior; *m*, third leg, posterior; *n*, fourth leg, posterior; and *o*, fifth pair of legs, posterior.

proximal to distal, on 5 endopodal segments.

First leg exopod (Figure 19*k*) with a relatively small outer spine on each segment. Coxa of second and third legs (Figures 19*l* and 19*m*) with a distinct notch on internal as well as external margin and patches of spinules on posterior surface. In each of second to fourth leg exopods (Figure 19*n*) all outer spines of approximately equal size and posterior sur-

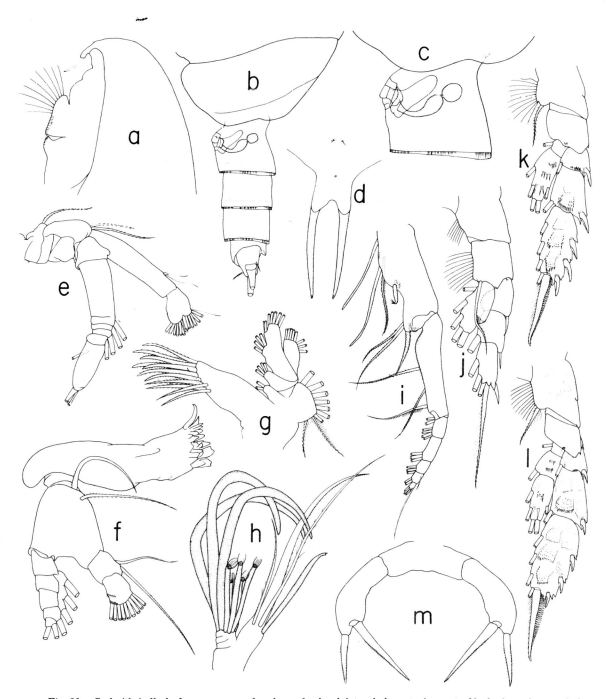

Fig. 20. *Scolecithricella hadrosoma*, n. sp., female: *a*, forehead, lateral; *b*, posterior part of body, lateral; *c*, genital segment, lateral; *d*, rostrum, anterior; *e*, antenna; *f*, mandible; *g*, maxillule; *h*, distal part of maxilla; *i*, maxilliped; *j*, first leg, anterior; *k*, second leg, posterior; *l*, third leg, posterior; and *m*, fifth pair of legs, posterior.

face densely armed with spinules. Endopods of first to fourth legs without spiniform distal process along external margin as found in *S. minor* or in *S. dentata*. Fifth pair of legs (Figure 19*o*) symmetrical, each leg 2-segmented, curved inward. Distal segment elongate, partially divided by a short line midway on inner side, with a large internal and a small distal spine. Internal spine shorter than segment and more than twice as long as distal spine.

Remarks. The specimens referred here to *Scolecithricella emarginata* showed a wide range of variation in body length (3.64–4.80 mm). However, since it was not possible to find morphological differences among individuals that could be used to

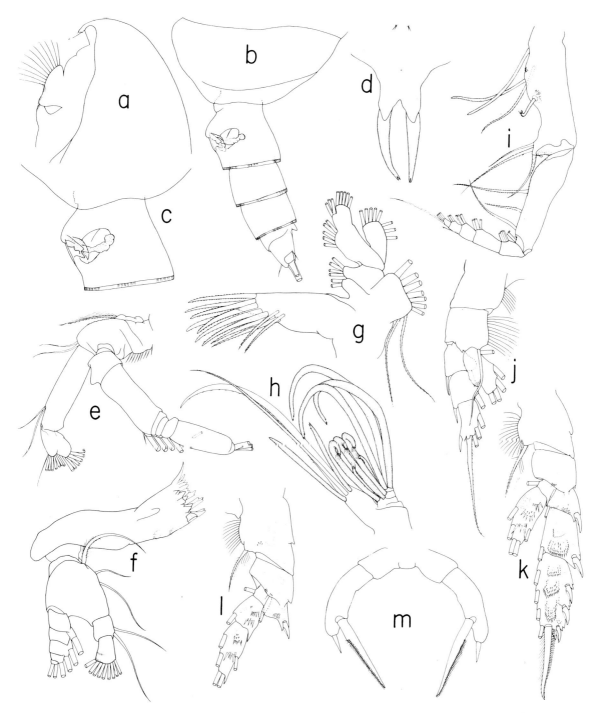

Fig. 21. *Scolecithricella obtusifrons* female: *a*, forehead, lateral; *b*, posterior part of body, lateral; *c*, genital segment, lateral; *d*, rostrum, anterior; *e*, antenna; *f*, mandible; *g*, maxillule; *h*, distal part of maxilla; *i*, maxilliped; *j*, first leg, anterior; *k*, second, leg, posterior; *l*, third leg without two distal exopodal segments, posterior; and *m*, fifth pair of legs, posterior.

separate them into subgroups, all specimens were considered here as belonging to a single species, although the size range displayed by them seemed to suggest the inclusion of more than one species.

Scolecithricella emarginata was originally described as *Scolecithrix emarginata* by Farran [1905] from a female 4.3 mm long taken off the west coast of Ireland and has been redescribed by With

[1915] (as *Scaphocalanus obtusifrons*), Sars [1925] (as *Amallothrix emarginata*), and Tanaka [1962] (as *Scolecithricella emarginata*). As has been summarized by Tanaka [1962] and Vervoort [1965], this species has been known to occur widely in deep waters of all the oceans. In the Atlantic Ocean the range extends from 65°27'N to 55°S. The body size of the species has been known to vary from 3.96 mm in the Bay of Biscay [Farran, 1926] to 4.73 mm in the Gulf of Guinea [Vervoort, 1965].

The specimens found in the present study seemed agreeable in all morphological details as well as in body size wtih the descriptions by Farran [1905], With [1915], Sars [1925], and Tanaka [1962]. They were also identical in all meristic details with specimens from the Gulf of Mexico (T. Park, unpublished data, 1977). Furthermore, they seemed undoubtedly identical with *Scolecithrix polaris*, originally described by Wolfenden [1911] from female specimens 3.5-4.0 mm long taken in the antarctic ice region. Its exact type locality was not given. Farran [1929] recorded *S. polaris* from 71°41'S, 166°47'W on the basis of four females 4.38 mm in body length. Vervoort [1957] reported *S. polaris* from eight stations between 61°44'S and 66°15'S and between 49°16'E and 178°29.5'E in the Indo-Pacific sector of the Antarctic. His female specimens were 4.28-4.41 mm long.

The male of *S. emarginata* has been described by With [1915] (as *Scaphocalanus obtusifrons*) and Tanaka [1962]. Wolfenden [1911] described the male of *S. polaris*, but, as was pointed out by Vervoort [1957], his male seems to belong to *Scaphocalanus*. In the present study, a total of 226 females were found from 38 stations distributed throughout the study area. All samples containing the species were taken from depths greater than 500 m. No males referable to *S. emarginata* were found during the study.

Scolecithricella hadrosoma, n. sp.
Fig. 20

Occurrence. The following station list shows the occurrence of *S. hadrosoma*, n. sp.:

Eltanin Cruise 17

Sta. 26, 2560-0 m, 2F (5.25-5.66 mm)

Eltanin Cruise 21

Sta. 198, 2972-0 m, 1F (5.00 mm)

Eltanin Cruise 22

Sta. 1528, 2452-0 m, 2F (5.50 mm)

Eltanin Cruise 32

Sta. 2133, 1829-0 m, 2F (5.25-5.33 mm)

Total: 7F

Female. Prosome length, 4.16-4.58 mm; body length, 5.00-5.66 mm. Similar in all morphological details to *S. emarginata* except characters noted below.

Body considerably larger than *S. emarginata*. Articulation suture between fourth and fifth metasomal segments almost complete, extending from dorsal side close to ventral edge (Figure 20b). Viewed laterally, genital field (Figure 20c) only slightly produced, so that ventral margin of genital segment posterior to genital opening almost straight instead of curved as in *S. emarginata*. Lateral skeletal plate of genital orifice narrowly elongated. Distal sac of spermathecal vesicle fully inflated into a large ball. Fifth leg (Figure 20m) 2-segmented, curved inward, with an internal and a distal spine as in *S. emarginata*, but distal spine almost 2/3 as long as internal spine. Distal segment without trace of partial division by a short line found in *S. emarginata*.

Remarks. Although it is very close to *S. emarginata*, *S. hadrosoma* is believed to be a valid new species because it is unmistakably distinguishable by its large body size and some minor but consistent differences in the genital segment and the fifth pair of legs.

Altogether, seven females were found during the present study. They were taken in samples collected obliquely down to depths exceeding 1829 m at four widely scattered stations. Apparently, the species is rare but has a wide range of distribution. The specific name *hadrosoma* refers to the large body size.

A type specimen selected from the specimens taken at station 26 of *Eltanin* cruise 17 has been deposited in the U.S. National Museum of Natural History. Holotype, USNM catalog no. 170766.

Scolecithricella obtusifrons (Sars, 1905)
Fig. 21

Amallophora obtusifrons Sars, 1905, p. 22.
Amallothrix obtusifrons; Sars, 1925, p. 179, pl. 50, figs. 1-16.—Wilson, 1942, p. 171, fig. 128; 1950, p.

162, pl. 4, figs. 1, 2.—Davis, 1949, p. 45.

? *Scolecithricella tydemani* A. Scott, 1909, p. 93, pl. 30, figs. 10–17.

? *Scolecithricella curticauda* A. Scott, 1909, p. 94, pl. 30, figs. 1–9.

Non *Scolecithricella obtusifrons;* **A. Scott, 1909, p. 92, pl. 31, figs. 1-9** (< *Scolecithricella emarginata*).

Non *Scaphocalanus obtusifrons*; With, 1915, p. 194, pl. 7, fig. 9, pl. 8, fig. 8, text figs. 60, 61 (< *Scolecithricella emarginata*).

Occurrence. The following station list shows the occurrence of *S. obtusifrons* (Sars, 1905):

Eltanin Cruise 17

Sta. 26, 2560–0 m, 1F (6.00 mm)

Eltanin Cruise 21

Sta. 213, 1050–0 m, 3F (6.16–6.35 mm)
Sta. 240, 2470–0 m, 1F (6.16 mm)

Eltanin Cruise 22

Sta. 1568, 2359–0 m, 3F (5.75–6.25 mm)

Eltanin Cruise 35

Sta. 2285, 1250–0 m, 1F (6.16 mm)

Atlantis II Cruise 31

Sta. RHB 1440, 1295–0 m, 4F (5.25–5.58 mm)

Total: 13F

Female. Prosome length, 4.41–5.33 mm; body length, 5.25–6.35 mm. Similar in habitus to *Scolecithricella emarginata* but differs from it in several characters as described below.

Articulation suture separating fourth and fifth metasomal segments almost complete (Figure 21*b*). Incurvation on distal margin of metasome relatively shallow. Ventral margin of genital segment posterior to genital opening not so strongly curved as in *S. emarginata*. Spermathecal vesicle (Figure 21*c*) not fully inflated and without a separate distal sac. Rostral rami (Figure 21*d*) less divergent, and sensory filaments thicker and shorter than in *S. emarginata*.

Antennule and antenna (Figure 21*e*) as in *S. emarginata*. Basis of mandible (Figure 21*f*) with 3 setae. Maxillule (Figure 21*g*) with 5 setae on basis and 3 + 5 setae on endopod. Fifth lobe of maxilla (Figure 21*h*) with 2 setae and 2 vermiform sensory filaments. Maxilliped (Figure 21*i*) and first to third legs (Figures 21*j*–21*l*) agree in all anatomical details with those of *S. emarginata*. Fifth leg (Figure 21*m*) 2-segmented, curved inward, with a large internal and a small distal spine. Internal spine slightly longer than distal segment and about 4 times as long as distal spine.

Remarks. *Scolecithricella obtusifrons* was originally described briefly as *Amallophora obtusifrons* by Sars [1905] from female specimens 5.60 mm long obtained in two deep tows at 45°30'N, 05°50'W and 36°17'N, 28°53'W, respectively, in the North Atlantic. Sars [1925] redescribed the species, including the male, as *Amallothrix obtusifrons* in detail with figures and added two more localities to its distribution (31°46'N, 25°00'W and 35°56'N, 08°00'W).

Scolecithricella obtusifrons has often been confused with *S. emarginata* (Farran, 1905), but according to the original descriptions they are clearly distinguishable by the body size and the meristic details of the maxillule. In the present study, the two species were fully described to provide more reliable diagnostic characters. Most useful in distinguishing the two species are the form of the spermatheca and the meristic details of the mandible, maxillule, and maxilla as described above.

With [1915] recorded *Scaphocalanus obtusifrons* from the Norwegian Sea upon female (4.4 mm long) and male (3.84 mm long) specimens. A Scott [1909] recorded *Scolecithricella obtusifrons* from the Malay Archipelago upon female specimens 4.3 mm long. These records by With [1915] and A. Scott [1909] have been referred to *Scolecithricella emarginata* by Sewell [1929].

A. Scott [1909] described *Scolecithricella tydemani* and *Scolecithricella curticauda* from female specimens 5.7 mm and 6.0 mm long, respectively, that were found in the same tow taken in the Malay Archipelago. Sars [1925] considered *S. tydemani* synonymous with *S. obtusifrons*. According to A. Scott [1909], *S. tydemani* and *S. curticauda* are the same in the details of the appendages except for the fifth leg, which is three segmented in the former and two segmented in the latter. In both species the maxilla has two vermiform sensory filaments on the fifth lobe, as is true in the specimens of *S. obtusifrons* examined in the present study. As the fifth legs are often variable, it is quite possible that both *S. tydemani* and *S. curticauda* are synonymous with *S. obtusifrons*.

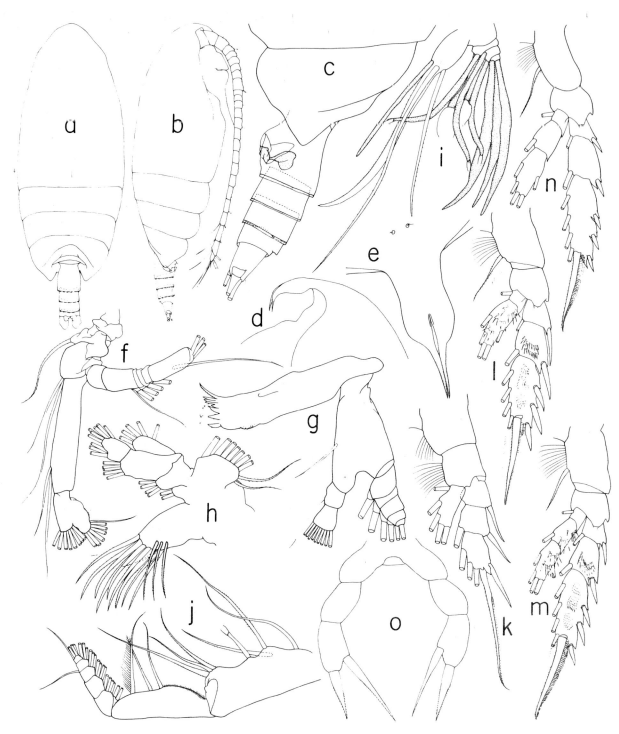

Fig. 22. *Scolecithricella altera* female: *a*, habitus, dorsal; *b*, habitus, lateral; *c*, posterior part of body, lateral; *d*, anterior end of head, lateral; *e*, rostrum, anterior; *f*, antenna; *g*, mandible; *h*, maxillule; *i*, distal part of maxilla; *j*, max-illiped; *k*, first leg, anterior; *l*, second leg, posterior; *m*, third leg, posterior; *n*, fourth leg, posterior; and *o*, fifth pair of legs, posterior.

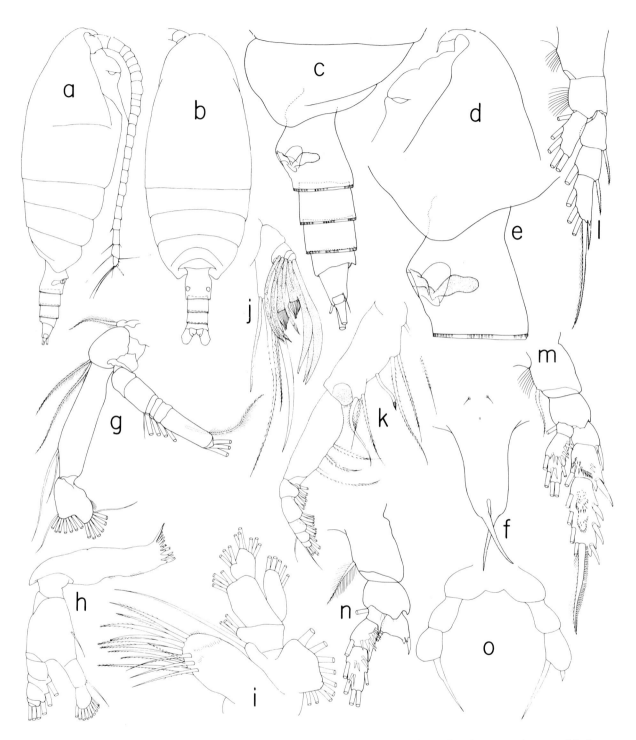

Fig. 23. *Scolecithricella vervoorti*, n. sp., female: *a*, habitus, lateral; *b*, habitus, dorsal; *c*, posterior part of body, lateral; *d*, forehead, lateral; *e*, genital segment, lateral; *f*, rostrum, anterior; *g*, antenna; *h*, mandible; *i*, maxillule; *j*, distal part of maxilla; *k*, maxilliped; *l*, first leg, anterior; *m*, second leg, posterior; *n*, third leg without two distal exopodal segments, posterior; and *o*, fifth pair legs, posterior.

Scolecithricella obtusifrons has also been recorded from the Pacific by Wilson [1942, 1950] and Davis [1949] as *Amallothrix obtusifrons*, but their descriptions are insufficient to prove the identity of their records.

In the present study, *S. obtusifrons* was represented by 13 females that were obtained from six widely distributed stations, of which two stations were from south of the Antarctic Convergence. The present finding represents the first record of the species in antarctic and subantarctic waters.

Scolecithricella altera (Farran, 1929)
Fig. 22

Amallophora altera Farran, 1929, p. 252, fig. 19.
Amallothrix altera; Roe, 1975, p. 330, fig. 18.
? *Scolecithricella altera*; Vervoort, 1965, p. 69, figs. 14–16.
Racovitzanus antarcticus; Vervoort, 1957, p. 97, figs. 81–84 (male only).
Non *Amallophora altera*; Vervoort, 1957, p. 94, figs. 77–79 (< *S. vervoorti*, n. sp.).—Bradford, 1973, p. 144, figs. 3*a*, *b*.

Occurrence. The following station list shows the occurrence of *S. altera* (**F**arran, 1929):

Eltanin Cruise 17

Sta. 59, 1251–0 m, 1F (2.80 mm)

Eltanin Cruise 46

Sta. 4, 1000 m, 1F (2.64 mm)
Sta. 8, 500–0 m, 2F (2.68 mm)
Sta. 9, 1000–0 m, 9F (2.56–2.72 mm)
Sta. 10, 1000–0 m, 5F (2.64–2.72 mm)
Sta. 11, 1000 m, 3F (2.64–2.72 mm)

Total: 21F

Female. Prosome length, 2.16–2.32 mm; body length, 2.56–2.80 mm. Body strongly built. Cephalosome and first metasomal segment fused (Figures 22*a* and 22*b*). Fourth and fifth metasomal segments partially separated by an articulation suture. Laterally, last metasomal segment produced distad as an angular projection covering anterior $1/2$ of genital segment.

Urosome about $1/4$ length of prosome. Genital segment (Figure 22*c*) nearly twice as long as second urosomal segment. Third urosomal segment slightly shorter than second. Laterally, genital segment with well-developed genital swelling. Lateral skeletal plate of genital orifice elongated and attenuated distad. Spermatheca with large vesicle extending as deep as lateral skeletal plate of genital orifice and expanding on posterior side into an additional pouch. Rostrum (Figure 22*e*) large, distally divided into 2 rami, each ramus tapering abruptly into a thin filament. Distal part of filament transformed into delicate sensory structure. Laterally, rostral filaments strongly curved backward (Figure 22*d*).

Antennule (Figure 22*b*) with 24 free segments, eighth and ninth of 25 segments being fused, and reaching close to distal end of prosome. In antenna (Figure 22*f*), endopod considerably longer than exopod. Second and third exopodal segments fused. Seta belonging to second segment relatively well developed. Seventh exopodal segment with a large middle seta in addition to 3 large distal setae. Mandible (Figure 22*g*) with well-developed masticatory blade. Basis and first endopodal segment each with a seta. Maxillule (Figure 22*h*) with 3 posterior and 10 distal setae on first inner lobe, 2 setae on second, and 4 setae on third; 5 setae on basis, 3 + 6 setae on endopod, 5 setae on exopod, and 9 setae on outer lobe. Maxilla (Figure 22*i*) with 3 setae on each of first 4 lobes, 3 setae and 1 vermiform sensory filament on fifth lobe, and 2 short and 3 long brush-form and 3 long vermiform sensory filaments on endopod. Coxa of maxilliped (Figure 22*j*) with 6 well-developed setae and 1 short brush-form sensory filament; endopod with 4, 4, 3, 3 + 1, and 4 setae on first to fifth segments.

First leg (Figure 22*k*) with 1-segmented endopod and 3-segmented exopod. Endopod without external lobe. Each exopodal segment with a long outer spine. Basis without inner seta. In second leg (Figure 22*l*), exopod with long outer spines. Second endopodal and second and third exopodal segments heavily armed with spines on posterior surface. In third leg (Figure 22*m*), exopod with outer spines of medium size. Two distal segments of both endopod and exopod armed with spines on posterior surface. Fourth leg (Figure 22*n*) without conspicuous spinous armature on posterior surface. Exopod with small outer spines. Fifth leg (Figure 22*o*) 3-segmented. Second and third segments about equal in length. Third segment with an inner and a distal spine. Distal spine about as long as segment and a little shorter than inner spine.

Remarks. *Scolecithricella altera* was originally described as *Amallophora altera* by Farran [1929] from two female specimens 2.67–2.76 mm long taken

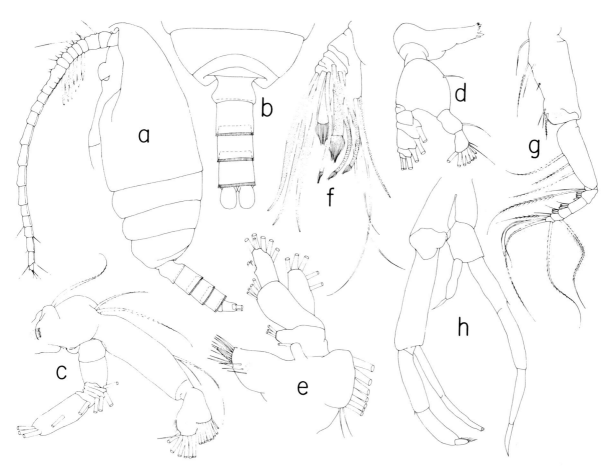

Fig. 24. *Scolecithricella vervoorti*, n. sp., male: *a*, habitus, lateral; *b*, posterior part of body, dorsal; *c*, antenna; *d*, mandible; *e*, maxilluble; *e*, maxillule; *f*, distal part of maxilla; *g*, maxilliped; and *h*, fifth pair of legs, posterior.

from 71°41′S, 166°47′W in the Ross Sea of the Antarctic. Roe [1975] found one female from 28°N, 14°W and three females from 18°N, 25°W in the northeastern Atlantic which he referred to *A. altera* Farran. Although they are considerably smaller in size (1.98–2.20 mm long), Roe's specimens agree in the meristic details of the appendages with Farran's [1929] as well as with those examined in the present study.

Vervoort [1957] reported two females from 66°11′S, 65°10′E and 63°51′S, 54°16′E, respectively, as belonging to *A. altera* Farran. As was pointed out by Roe [1975], Vervoort's specimens are significantly different from Farran's. During the present study, specimens identical with Vervoort's *A. altera* were found, and they are described below as a new species. Vervoort [1965] reported *Scolecithricella altera* from the Gulf of Guinea on the basis of a single female (3.00 mm long) which was, however, not only slightly

larger than, but also different morphologically from, the antarctic specimens described by Farran [1929] and those found in the present study in having a well-developed inner seta on the basis of the first leg. Bradford [1973] redescribed *A. altera*, showing a greatly enlarged brush-form sensory filament on the maxilla endopod. As was pointed out already by Roe [1975], Bradford's [1973] species is not referable to *A. altera*.

Vervoort [1957] described a male (2.34 mm long) from the Antarctic as belonging to *Racovitzanus antarcticus*. Although it is similar to the female of *R. antarcticus* in body size, his male is obviously different from it in having an outer spine on the first exopodal segment of the first leg. Upon close examination of the description, Vervoort's male was found to be agreeable in body size as well as in all anatomical details with the female of *S. altera* as redescribed here. It is therefore believed that Vervoort's male is the hitherto unknown male of *S. altera.*

Scolecithricella altera was represented in the present study by 21 females taken from five stations in antarctic waters and one station on the Antarctic Convergence.

Scolecithricella vervoorti, n. sp.
Figs. 23 and 24

Amallophora altera; Vervoort, 1957, p. 94, figs. 77–79.

Occurrence. The following station list shows the occurrence of *S. vervoorti*, n. sp.:

Eltanin Cruise 17

Sta. 41, 625–0 m, 2F (2.64–2.76 mm);
1M (2.48 mm)
Sta. 52, 1052–0 m, 1F (2.64 mm)

Eltanin Cruise 46

Sta. 5, 500 m, 13F (2.56–2.64 mm)
1000 m, 4F (2.56–2.72 mm)
Sta. 6, 500 m, 3F (2.56–2.60 mm)
1000 m, 2F (2.52–2.64 mm)
Sta. 7, 500 m, 9F (2.56–2.76 mm)
Sta. 9, 1000–0 m, 6F (2.52–2.76 mm)
Sta. 10, 1000–0 m, 2F (2.56–2.68 mm)
Sta. 11, 1000 m, 1F (2.60 mm)
Sta. 15, 1000 m, 1F (2.72 mm)
Sta. 16, 500 m, 5F (2.56–2.60 mm)
Sta. 17, 500 m, 1F (2.56 mm)

Total: 50F and 1M

Female. Prosome length, 2.00–2.32 mm; body length, 2.52–2.76 mm. Very close morphologically to *Scolecithricella altera* but readily distinguishable from it by a number of characters as described below. Posterolateral corners of metasome produced distad, covering anterior 1/3 of genital segment. Viewed laterally, posterior corner of metasome round (Figures 23c and 23e). Lateral skeletal plate of genital orifice short and broadly round. Spermathecal vesicle large but without a pouchlike expansion on posterior side as in *S. altera.* Viewed anteriorly, rostrum (Figure 23f) with an oblong basal plate.

Antennule (Figure 23a), antenna (Figure 23g), and mandible (Figure 23h) agree in all meristic details with those of *S. altera.* Maxillule (Figure 23i)

with 1 and 3 setae on second and third inner lobes, respectively, and 4 setae on basis; setation otherwise same as in *S. altera.* Maxilla (Figure 23j) and maxilliped (Figure 23k) similar to those of *S. altera* except that fourth endopodal segment of maxilliped bears 2 + 1 setae instead of 3 + 1. First to third legs (Figures 23l–23n) similar to those of *S. altera.* Fifth leg (Figure 23o) also similar to that of *S. altera*, but distal spine very small and often missing.

Male. Prosome length, 1.92 mm; body length, 2.48 mm. Cephalosome and first metasomal segment fused. Fourth and fifth metasomal segments partially separated by articulation suture (Figures 24a and 24b). Urosome about 1/3 as long as prosome, with second to fourth segments similar in length and wider than long. Rostrum mutilated.

Antennule (Figure 24a) reaching approximately distal end of prosome, with eighth to twelfth segments partially or fully fused. Twentieth and twenty-first segments fused in right antennula but separate in left. Antenna (Figure 24c) with well-developed seta on second exopodal segment. Mandible (Figure 24d) similar in meristic details to that of female, but masticatory blade considerably reduced. Maxillule (Figure 24e) with inner lobes much reduced but agrees in meristic details with that of female except that endopod with 2 + 6 setae instead of 3 + 6. Maxilla (Figure 24f) similar to that of female, but setae on first to fifth lobes slightly reduced in size. Maxilliped (Figure 24g) agrees in setation with that of female, but setae on coxa poorly developed, and distal and external setae on endopod better developed than in female. First leg as in female in all details. Second to fourth legs were mutilated. Right fifth leg (Figure 24h) with moderately developed endopod and slender exopod extending to nearly same length as left exopod. Left fifth leg with relatively long endopod, only slightly shorter than exopod.

Remarks. Scolecithricella vervoorti is closely related to *Scolecithricella altera* (Farran, 1929) but clearly distinct from it in the form of the posterior metasomal corner and genital segment, in the setation of the maxillule and maxilliped, and in the form of the fifth legs.

The male described here is believed to belong to this species mainly because of its agreement in body size and meristic details of the appendages with the female. The fifth pair of legs is very similar to that of a *Racovitzanus antarcticus* male as described by Vervoort [1957], which is believed in this study to be referable to *Scolecithricella altera* as described above.

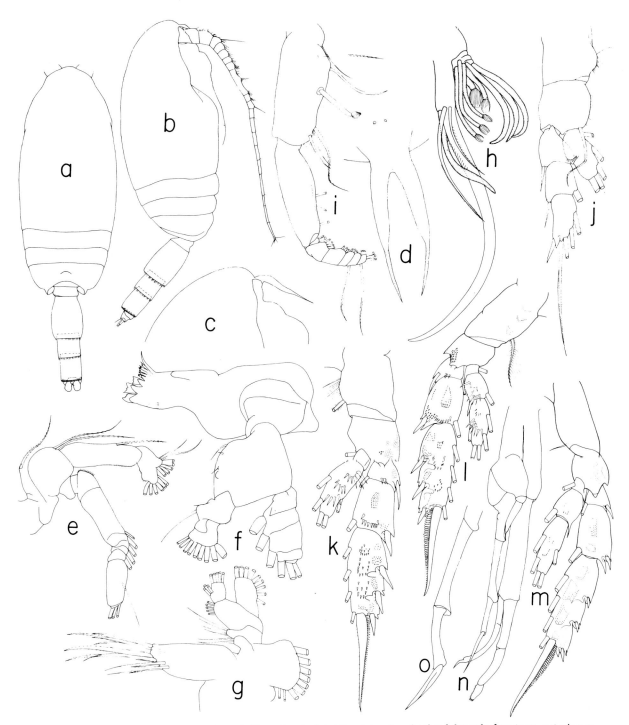

Fig. 25. *Scolecithricella* sp. 1 male: *a*, habitus, dorsal; *b*, habitus, lateral; *c*, forehead, lateral; *d*, rostrum, anterior; *e*, antenna; *f*, mandible; *g*, maxillule; *h*, distal part of maxilla; *i*, maxilliped; *j*, first leg, anterior; *k*, second leg, posterior; *l*, third leg, posterior; *m*, fourth leg, posterior; *n*, fifth pair of legs, anterior; and *o*, right fifth leg exopod, anterior.

This new species is named after Willem Vervoort of Leiden, the Netherlands, who [Vervoort, 1957] has recorded the species from the Antarctic as belonging to *Scolecithricella altera*.

Scolecithricella vervoorti seemed fairly common in the Antarctic. Altogether, 50 females and one male were found in the present study. They were taken at 11 stations in waters on and south of the Antarctic

Convergence. A type specimen selected from the specimens taken at station 7 on *Eltanin* cruise 46 has been deposited in the U.S. National Museum of Natural History. Holotype female, USNM catalog no. 170767.

Scolecithricella sp. 1
Fig. 25

Occurrence. The following station list shows the occurrence of *S.* sp. 1:

Eltanin Cruise 17

Sta. 26, 2560–0 m, 1M (4.91 mm)
Sta. 59, 1251–0 m, 3M (4.75–5.00 mm)
Sta. 62, 1251–0 m, 1M (5.00 mm)

Eltanin Cruise 23

Sta. 1685, 2250–0 m, 3M (4.83 mm)

Eltanin Crusie 33

Sta. 2174, 1830–0 m, 2M (5.25 mm)

Eltanin Cruise 35

Sta. 2293, 1300–0 m, 2M (5.00–5.08 mm)

Eltanin Cruise 46

Sta. 9, 1000–0 m, 1M (4.91 mm)
Sta. 10, 1000–0 m, 2M (4.68–4.75 mm)
Sta. 11, 1000 m, 2M (4.75 mm)

Total: 17M

Male. Prosome length, 3.36–3.58 mm; body length, 4.68–5.25 mm. Similar morphologically to males of *S. dentipes, S. pseudopropinqua,* and *S. valida* described above but distinguishable from them mainly by large body size and some details of fifth pair of legs as described below.

Urosome nearly ¹/₂ as long as prosome (Figures 25a and 25b). Second urosomal segment almost equal to combined lengths of 2 following segments. Rostrum (Figure 25d) with 2 strong rami, each terminating in a delicate sensory filament. Antennule (Figure 25b) reaching about distal end of prosome, with eighth to twelfth segments fused. Twentieth and twenty-first segments fused in right antennule but separated in left.

Antenna (Figure 25e), mandible (Figure 25f), maxillule (Figure 25g), maxilla (Figure 25h), maxilliped (Figure 25i) as in *S. dentipes* male except that basis of mandible with 3 minute setae instead of 2, and, as in *S. pseudopropinqua* and *S. valida,* 3 distal

setae of maxilliped coxa well developed.

First to fourth legs (Figures 25j–25m) similar to those of *S. dentipes* male. In right fifth leg (Figures 25n and 25o), first and second exopodal segments separated by a suture. Second exopodal segment reaching distal end of left basis. Third exopodal segment relatively short, nearly straight, bearing a long terminal spine about as long as segment itself. Endopod short of reaching distal end of first exopodal segment. Left fifth leg with 3-segmented exopod and 1-segmented endopod. Endopod about as long as combined lengths of first 2 exopodal segments. Third exopodal segment shorter than ¹/₂ length of second.

Remarks. In the habitus and the meristic details of the appendages this male is closely related to the *S. dentipes* group discussed below under the heading of general remarks. However, none of the females of the *S. dentipes* group found in the present study were large enough to match this male.

Of all females found in the present study, that of *S. emarginata* was most similar in body size to this male. Although the male of *S. emarginata* was not found in the study, this male cannot be referred to this species because of a number of obvious morphological differences, namely, the shape of the rostrum and the number of setae on the mandible, maxillule, and maxilliped. Furthermore, the male described for *S. emarginata* by With [1915] (as *S. obtusifrons*) and Tanaka [1962] is significantly different from the present male.

In the present study, the male was represented by 17 specimens found at nine stations exclusively in antarctic waters south of the Antarctic Convergence.

Scolecithricella sp. 2
Fig. 26

Occurrence. The following station list shows the occurrence of *S.* sp. 2:

Eltanin Cruise 21

Sta. 264, 1230–0 m, 1M (3.16 mm)

Eltanin Cruise 26

Sta. 1825, 1625–0 m, 1M (3.12 mm)
Sta. 1842, 1350–0 m, 1M

Eltanin Cruise 35

Sta. 2285, 1250–0 m, 1M (3.00 mm)

Total: 4M

Male. Prosome length, 2.16–2.28 mm; body

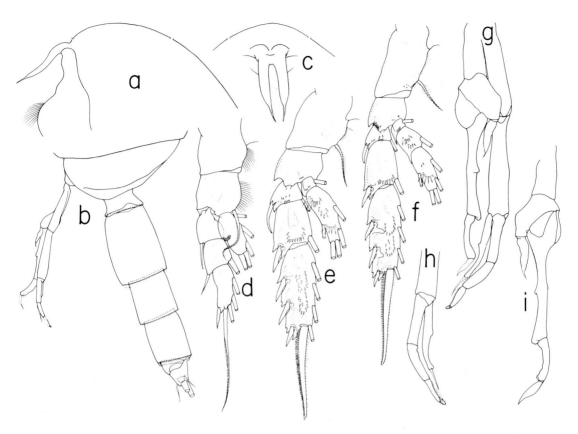

Fig. 26. *Scolecithricella* sp. 2 male: *a*, forehead, lateral; *b*, posterior part of body, lateral; *c*, rostrum, anterior; *d*, first leg, anterior; *e*, second leg, posterior; *f*, third leg, posterior; *g*, fifth pair of legs, anterior; *h*, distal part of left fifth leg, posterior; and *i*, right fifth leg, posterior.

length, 3.00–3.16 mm. Similar in habitus to *Scolecithricella* sp. 1 described above, but second urosomal segment about 1.5 times as long as third or fourth urosomal segment (Figure 26*b*). Cephalosomal appendages from antennule to maxilliped agree in all meristic details with *Scolecithricella* sp. 1 except that maxillule with 8 setae, instead of 9, on exopod. First to third legs (Figures 26*d*–26*f*) similar to those of *Scolecithricella* sp. 1, but first exopodal spine of second leg long and strongly curved and posterior surfaces of all legs densely covered by spinules.

In right fifth leg (Figures 26*g*–26*i*), first and second exopodal segments almost completely fused. Second exopodal segment reaching about distal end of left basis. Third exopodal segment slightly curved inward, reaching distal end of first exopodal segment of left leg. Terminal spine about ²/₃ as long as segment. Endopod about ¹/₃ as long as first exopodal segment. Left fifth leg with 3-segmented exopod and 1-segmented endopod. Third exopodal segment about ²/₃ as long as second. Endopod considerably shorter than combined lengths of first and second exopodal

segments and with a long terminal spine.

Remarks. This male is the most similar morphologically to the *Scolecithricella* sp. 1 male and the next most similar to males of *S. dentipes, S. pseudopropinqua,* and *S. valida* described above. Of all the females found in the present study, the *S. pseudopropinqua* female is the closest in body size to this male. However, the *S. pseudopropinqua* female agrees more closely in body size and anatomical details, including the pattern of the spinous armature of the legs, with the male referred to that species than with this male.

The male was represented in the present study by a total of four specimens found at four stations, all being in subantarctic waters north of the Antarctic Convergence.

Scolecithricella sp. 3
Fig. 27

Occurrence. The following station list shows the occurrence of *S.* sp. 3:

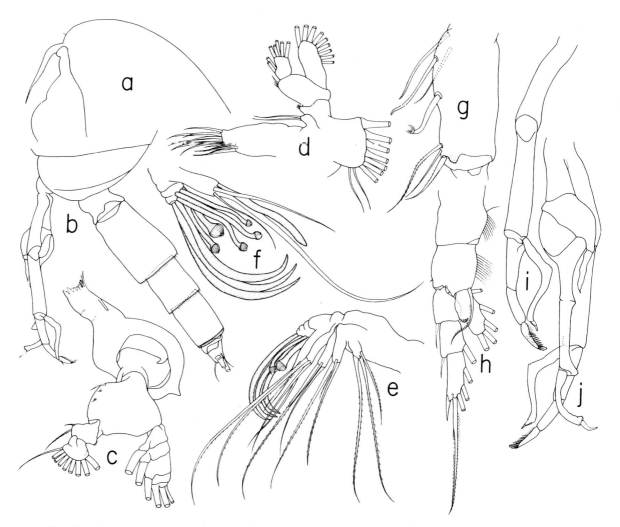

Fig. 27. *Scolecithricella* sp. 3 male: *a*, forehead, lateral; *b*, posterior part of body, lateral; *c*, mandible; *d*, maxillule; *e*, maxilla; *f*, distal part of maxilla; *g*, coxa of maxilliped; *h*, first leg, anterior; *i*, left fifth leg, posterior; and *j*, fifth pair of legs, right side.

Eltanin Cruise 23

Sta. 1700, 1275–0 m, 1 M (3.80 mm)

Male. Prosome length, 2.64 mm; body length, 3.80 mm. Similar in habitus to *S. dentipes*. Second urosomal segment longer than third or fourth by ²/₅ its length (Figure 27*b*). Rostrum, antennule, and antenna as in *S. dentipes*. Basis of mandible (Figure 27*c*) with 3 minute setae and without a conical process. Maxillule (Figure 27*d*) with 2 posterior and 10 distal setae on first inner lobe, 2 setae on second, and 3 setae on third; 3 setae on basis, 2 + 5 setae on endopod, 9 setae on exopod, and 9 setae on outer lobe. Setae on third inner lobe and basis greatly reduced.

Maxilla (Figure 27*f*) with same number of setae and sensory filaments as in *S. dentipes*, but setae are poorly developed. Maxilliped (Figure 27*g*) also agrees in setation with *S. dentipes*, but 3 distal setae of coxa well developed as in *S. pseudopropinqua*. First leg (Figure 27*h*) as in *S. dentipes*. Second to fourth legs mutilated.

In right fifth leg (Figures 27*i* and 27*j*), first and second exopodal segments partially separated by an incomplete suture. Second exopodal segment with a large distal process and reaching close to middle of first exopodal segment of left leg. Third exopodal segment strongly curved inward and with a small terminal spine. Endopod short of reaching distal end of first exopodal segment. Left fifth leg with 3-segmented exopod and 1-segmented endopod. Endopod strongly bent like an elbow and extending beyond distal end of second exopodal segment.

Remarks. This male differs from all other *Scolecithricella* males found in the study in having greatly reduced setae on the third inner lobe and basis of the maxillule and very weakly developed setae on the maxillar lobes. However, in view of the similarity exhibited by the rostrum and fifth pair of legs this male seems to be closely related to the *S. dentipes* group discussed below.

GENERAL REMARKS

According to the anatomical details the *Scolecithricella* species found in present study can be divided into four groups, as discussed below.

The minor group. The *minor* group consists of *S. minor*, *S. profunda*, *S. vittata*, *S. dentata*, and *S. schizosoma.* The rostrum consists of a short basal plate which is divided into two strong rami. Only the distal part of each ramus is transformed into a sensory filament of varying size. In the antenna the exopod is considerably longer than the endopod. The basis and the first endopodal segment of the mandible have a single seta each. The first inner lobe of the maxillule has one or two posterior setae, and the third inner lobe has three setae. The first exopodal segment of the first leg has no external spine. The second to fourth leg coxae have no notches on the internal and external margins. In the female the fifth leg is one segmented and leaflike. In the male the second to fourth urosomal segments are longer than they are wide. The right fifth leg has a small endopod.

The dentipes group. The *dentipes* group consists of *S. dentipes*, *S. parafalcifer*, *S. pseudopropinqua*, *S. valida*, *S.* sp. 1, and *S.* sp. 2. The rostrum and the antenna are the same as those in the *minor* group. The mandibular basis has three setae in the female, the middle one of which is very small, and two or three small setae and a conical process in the male. The first endopodal segment of the mandible has two setae. The first inner lobe of the maxillule has two posterior setae, and the third inner lobe has four setae. The first exopodal segment of the first leg has an external spine. The second to fourth leg coxae have no notches on the internal and external margins. The female fifth leg is two segmented and elongated. In the male the second to fourth urosomal segments are longer than they are wide; the second is significantly longer than the others. The right fifth leg endopod is relatively well developed.

The ovata group. The *ovata* group consists of *S. ovata*, *S. cenotelis*, *S. emarginata*, *S. hadrosoma*, and *S. obtusifrons.* The rostrum consists of a short basal plate which is divided into two rami, as it is in the *minor* and *dentipes* groups. However, the rostral ramus is almost entirely transformed into a sensory filament. The antenna is like that in the *minor* and *dentipes* groups. The basis of the mandible has two or three long setae, and the first endopodal segment has two setae. In the maxillule the first inner lobe has two posterior setae, and the third has three or four setae. The first leg is like that in the *dentipes* group. The second and fourth legs have a conspicuous notch on the internal and external margins of the coxa. The female fifth leg is two segmented, and the distal segment is leaflike or elongated. The male has not been found in the study.

The altera group. The *altera* group consists of *S. altera* and *S. vervoorti.* The rostrum consists of a large, elongate basal plate which is divided distally into two rami. The distal part of the ramus is transformed into a sensory filament. In the antenna the exopod is considerably shorter than the endopod. The basis and the first endopodal segment of the mandible have a single seta each. The first inner lobe of the maxillule has three posterior setae, and the third inner lobe has three or four setae. In the first leg the basis is devoid of setae, and the endopod has no external lobe. The second to fourth legs have no notches on the internal and external margins of the coxa. The female fifth leg is three segmented and elongated. In the male the second to fourth urosomal segments are wider than they are long. The basis of the right fifth leg is not so greatly enlarged as that in the other groups.

Of the four groups described above, the *altera* group is significantly different from the others in having a large rostral plate, a short antennal exopod, and different meristic characters of the mandible, maxillule, and first leg. These differences seem to warrant the distinction of the group as a separate taxon. There is a male species (*S.* sp. 3) that has not been assigned to any of the four groups. This male is similar to the *dentipes* group in all characters except that the mandible has no conical process on the basis, the setae on the third inner lobe and the basis of the maxillule are greatly reduced, and the setae on the maxillary lobes are only weakly developed.

Acknowledgments. This study was carried out under the Cooperative Systematics Research Program of the Smithsonian Institution, supported by National Science Foundation grant OPP71-04058 A03. I am indebted to Betty J. Landrum, Director, Smithsonian Oceanographic Sorting Center, for her encouragement and support. Most of the materials reported herein were collected on the USNS *Eltanin* and processed at the Smithsonian Oceanographic Sorting Center. I am grateful to Frank D. Ferrari, Supervisor for Plankton, for making the materials available to the study. The samples collected by Richard H. Backus on R/V *Atlantis II* cruise 31 were obtained through J. E. Craddock and R. L. Haedrich of the Woods Hole Oceanographic Institution.

REFERENCES

Bradford, J. M.
  1972 Systematics and ecology of New Zealand central east coast plankton sampled at Kaikoura, Bull. 207, pp. 1–87, figs. 1–63, N. Z. Dep. of Sci. and Ind. Res., Wellington, New Zealand.
  1973 Revision of family and some generic definitions in the Phaennidae and Scolecithricidae (Copepoda: Calanoida), N. Z. J. Mar. Freshwater Res., *7*(1–2), 133–152, figs. 1–4.

Brady, G. S.
  1883 Report on the Copepoda collected by H. M. S. *Challenger* during the years 1873–76, Rep. Scient. Results Voyage H. M. S. *Challenger* 1873–76, *8*(23), 1–142, pls. 1–55.

Brodsky, K. A.
  1950 Calanoida of the far eastern seas and polar basin of the USSR (in Russian), Opred. Faune SSSR, no. 35, 1–442, figs. 1–306.

Davis, C. C.
  1949 The pelagic Copepoda of the northeastern Pacific Ocean, Univ. Wash. Publs Biol., No. 14, 1–117, pls. 1–15.

Farran, G. P.
  1905 Report on the Copepoda of the Atlantic slope off Counties Mayo and Galway, Annual Report on Fisheries, Ireland, 1902–03, part 2, append. 2, pp. 23–52, pls. 3–13, Dep. of Agr. and Tech. Instr. for Ireland, Fish. Br., Ireland.
  1908 Second report on the Copepoda of the Irish Atlantic Slope, Scient. Invest. Fish. Brch. Ire., no. 2, 104 pp., 11 pls.
  1926 Biscayan plankton collected during a cruise of H. M. S. *Research*, 1900, XIV, The Copepoda, J. Linn. Soc., *36*, 219–310, text figs. 1, 2, pls. 5–10.
  1929 Crustacea, X, Copepoda, Nat. Hist. Rep. Br. Antarct. Terra Nova Exped., *8*(3), 203–306, figs. 1–37, maps 1–4.
  1936 Copepoda, Scient. Rep. Gt Barrier Reef Exped., *5*(3), 73–142, text figs. 1–30.

Giesbrecht, W.
  1888 Elenco dei copepodi pelagici raccolti dal tenente di vascello Gaetano Chierchia durante il viaggio della R. Corvetta 'Vettor Pisani' negli anni 1882–1885, e dal tenente di vascello Francesco Orsini nel Mar Rosso, nel 1884, Atti Accad. Naz. Lincei Rc., Ser. 4, *4*(2), 284–287, 330–338.
  1892 Systematik und Faunistik der pelagischen Copepoden des Golfes von Neapel und der angrenzenden Meeresabschnitte, Fauna Flora Golf. Neapel, no. 19, 831 pp., 54 pls.
  1902 Copepoden, Result. Voyage S.Y. *Belgica*, 49 pp., 13 pls.

Grice, G. D.
  1962 Calanoid copepods from equatorial waters of the Pacific Ocean, Fishery Bull. Fish Wildl. Serv. U.S., *61*, 167–246, fig. 1, pls. 1–34.

Minoda, T.
  1971 Pelagic Copepoda in the Bering Sea and the north-

western North Pacific with special reference to their vertical distribution, Mem. Fac. Fish. Hokkaido Univ., *18*(1–2), 1–74, figs. 1–33, pls. 1–5, appends. 1–5.

Mori, T.
  1937 *The Pelagic Copepoda From the Neighbouring Waters of Japan*, 150 pp., 80 pls., Yokendo, Tokyo.

Owre, H. B., and M. Foyo
  1967 Copepods of the Florida Current, Fauna Caribaea, no. 1, 137 pp., 900 figs., Inst. of Mar. Sci., Univ. of Miami, Miami, Fla.

Park, T.
  1968 Calanoid copepods from the central North Pacific Ocean, Fishery Bull. Fish Wildl. Serv. U.S., *66*(3), 527–572, pls. 1–13.
  1978 Calanoid copepods (Aetideidae, Euchaetidae) from antarctic and subantarctic waters, in *Biology of the Antarctic Seas VII, Antarctic Res. Ser.*, vol. 27, paper 4, edited by D. L. Pawson, pp. 91–290, AGU, Washington, D. C.

Roe, H. S. J.
  1975 Some new and rare species of calanoid copepods from the northeastern Atlantic, Bull. Br. Mus. Nat. Hist., *28*(7), 297–372, figs. 1–33.

Rose, M.
  1933 Copépodes pélagiques, Faune Fr., *26*, 1–374, figs. 1–456, pls. 1–19.
  1942 Les Scolecithricidae (Copépodes pélagiques) de la Baie d'Alger, Annls Inst. Oceanogr. Monaco, *21*(3), 113–170, figs. 1–58.

Sars, G. O.
  1902 Copepoda Calanoida, Acct Crustacea Norway, *4*(5–6), 49–72, pls. 33–48.
  1905 Liste préliminaire des Calanoidés recueillis pendant les campagnes de S.A.S. le Prince Albert de Monaco, 1, Bull. Inst. Oceanogr. Monaco, no. 26, 1–22.
  1920 Calanoidés recueillis pendant les campagnes de S.A.S. le Prince de Monaco, Bull. Inst. Oceanogr. Monaco, no. 377, 1–20.
  1925 Copépodes particulièrement bathypélagiques provenant des campagnes scientifiques du Prince Albert Ier de Monaco, Result. Camp. Scient. Prince Albert I, no. 69, 408 pp., 127 pls.

Scott, A.
  1909 The Copepoda of the *Siboga* expedition, 1, Free-swimming, littoral and semi-parasitic Copepoda, *Siboga* Exped., no. 29a, 323 pp., 69 pls.

Scott, T.
  1894 Report on Entomostraca from the Gulf of Guinea, Trans. Lin. Soc. Lond., Zool. Ser., *6*(1), 1–161, pls. 1–15.

Sewell, R. B. S.
  1929 The Copepoda of Indian seas: Calanoida, Mem. Indian Mus., *10*, 1–221, text figs. 1–81.

Tanaka, O.
  1962 The pelagic copepods of the Izu region, middle Japan, systematic account, VIII, Family Scolecithricidae (part 2), Publs Seto Mar. Biol. Lab., *10*(1), 35–90, figs. 127–151.

Vervoort, W.
  1951 Plankton copepods from the Atlantic sector of the Antarctic, Verh. K. Ned. Aked. Wet., sect. 2, *47*(4), 1–156, figs. 1–82.
  1957 Copepods from antarctic and sub-antarctic plankton samples, Rep. B.A.N.Z. Antarct. Res. Exped., Ser. B, *3*, 1–160, 138 figs.
  1965 Pelagic Copepoda, II, Copepoda Calanoida of the families Phaennidae up to and including Acartiidae, containing the description of a new species of Aetideidae, Atlantide Rep., no. 8, 9–216, figs. 1–41.

Von Vaupel-Klein, J. C.
  1970 Notes on a small collection of calanoid copepods from the northeastern Pacific, including the description of a new species of *Undinella* (fam. Tharybidae), Zool. Verh. Leiden, no. 110, 3–43, figs. 1–15.

Wilson, C. B.
  1932 The copepods of the Woods Hole region, Massachusetts, Bull. U.S. Natn. Mus., no. 158, 635 pp., 316 figs., 41 pls.
  1942 The copepods of the plankton gathered during the last cruise of the *Carnegie*, Publs. Carnegie Instn, no. 536, 1–237, figs. 1–136.
  1950 Copepods gathered by the United States fisheries steamer 'Albatross' from 1887 to 1909, chiefly in the Pacific Ocean, Bull. U.S. Natn. Mus., *14*(14), 141–441, pls. 2–36.

With, C.
  1915 Copepoda, 1, Calanoida Amphascandria, Dan. Ingolf Exped., *3*(4), 1–260, text figs. 1–79, pls. 1–8.

Wolfenden, R. N.
  1911 Die marinen Copepoden der Deutschen Südpolar-Expedition 1901–1903, II, Die pelagischen Copepoden der Westwinddrift und des südlichen Eismeers, Dt. Sudpol. Exped., *12*(4), 181–380, text fig. 1–82, pls. 22–41.

The Antarctic Research Series

Each paper submitted to the ARS is thoroughly reviewed by one or more recognized authorities in that discipline. Papers accepted for publication have met the high scientific and publication standards established by the board of associate editors. The members of the current board include Charles R. Bentley, chairman, Robert H. Eather, Louis S. Kornicker, Dennis E. Hayes, and Bruce Parker. Fred G. Alberts, Secretary to the U.S. Advisory Committee on Antarctic Names, gives valuable assistance in verifying place names, locations, and maps.

To get papers into circulation as quickly and economically as possible, a new format has been adopted for the Antarctic Research Series. Papers that have completed the review/revision cycle are assigned by the board of associate editors to a volume according to subject matter and are printed and issued individually as soon as production schedules allow. Several topical volumes may be in process at any time, and the release of an individual paper does not have to wait until all papers proposed for a given volume are ready. The individual papers are issued under soft cover, and pages are numbered consecutively within each volume so that they can be collected and bound together after the last paper is released. At the completion of a volume, subscribers with standing orders will be sent the title page, table of contents, and other front matter for the volume. Purchasers will be responsible for having the volume bound if they choose to do so. Entire volumes will be available on microfiche. Individuals interested in publishing in the the series should write to

Chairman
Antarctic Research Series Board of Associate Editors
American Geophysical Union
2000 Florida Avenue, N.W.
Washington, D. C. 20009